Infinite Potential

ALSO BY LOTHAR SCHÄFER

In Search of Divine Reality (1997)

Infinite Potential

What Quantum Physics Reveals About How We Should Live

LOTHAR SCHÄFER

Foreword by Deepak Chopra

DEEPAK
CHOPRA
BOOKS

Published in the United States by Deepak Chopra Books, an imprint of the
Crown Publishing Group, a division of Random House, Inc., New York.
www.crownpublishing.com

DEEPAK CHOPRA BOOKS and colophon
are trademarks of Random House, Inc.

Cataloging-in-Publication data is on file with the Library of Congress.

ISBN 978-0-307-98595-8
eISBN 978-0-307-98596-5

Printed in the United States of America

Book design by Lauren Dong
Jacket design by Michael Nagin
Jacket photography © Traumlichtfabrik/Getty Images

10 9 8 7 6 5 4 3 2 1

First Edition

To Gabriele,
friend of a lifetime,
and our loving children

Contents

Foreword by Deepak Chopra   xi

Preface: The Physics of Enlightenment   xix

Introduction
YOUR COSMIC POTENTIAL: BEING PART OF THE
UNIVERSE   *1*

PART ONE: THE NATURE OF REALITY

Chapter 1
MATERIALISM IS WRONG: THE BASIS OF THE MATERIAL
WORLD IS NONMATERIAL   *33*

Chapter 2
YOUR POTENTIAL IS REAL EVEN THOUGH YOU
CAN'T SEE IT: HOW VIRTUAL STATES ACT ON THE
VISIBLE WORLD   *54*

Chapter 3
WE ARE ALL CONNECTED: REALITY AS INDIVISIBLE
WHOLENESS   *75*

Chapter 4
Consciousness: A Cosmic Property   *95*

PART TWO: LIVING IN THIS REALITY

Chapter 5
Darwin Was Wrong: Evolution Needs Quantum
Selection and Cooperation   *127*

Chapter 6
World Ethos: Living in Harmony with the
Order of the Universe   *164*

Chapter 7
Your Evolving Mind: Integrative Consciousness and
a Leap into a New Human Species   *195*

Acknowledgments   *219*

Appendix for Chapter 1
On Single-Particle Interference and the Concept of
Potentiality Waves   *221*

Appendix for Chapter 2
How the Nonempirical Part of Reality Is Discovered in the
Virtual States of Atoms and Molecules   *245*

Notes   *271*
Sources   *287*
Index   *297*

Foreword

Deepak Chopra

*I*deas turn into high drama when they walk on stage and change people's lives. In physics, traditionally honored as "the queen of the sciences," this happens very rarely. No queen is more aloof or enigmatic. When Einstein discovered relativity, it was said (by the English philosopher Bertrand Russell) that only three people in the world understood it. Russell's arithmetic wasn't astronomically wrong if you were counting people outside physics. Relativity shook the foundations of three major concepts in the field: time, space, and gravity. The cosmos would never be the same—but people's everyday lives would.

Lothar Schäfer plunks a colorless word on stage—*potential*—which is as mundane as *relativity*. It's also the seed of an entirely new universe, however, one that Einstein could never accept. The new universe outlined here is conscious. We are thinking creatures because the cosmos thinks. We breathe with life because the primal ingredients of life are embedded in the fabric of Nature. Not just the elements that evolved into organic chemicals. The process of life itself is basic to the universe and has invisibly guided it since the Big Bang. A fan of James Joyce's writing once gushed that he used so many words. He replied that it wasn't the words, it was where you put them that matters. The universe has taken its vocabulary of atoms and molecules and arranged them precisely to give rise to human DNA and the human mind. This book tells how and why that happened. We are the expressed potential of creation, the way that Da Vinci's *Mona Lisa* is the expressed potential of his genius.

But it's not the science that I want to emphasize first but its human dimensions. No one, including Einstein himself, has had the nerve to subtitle a book, as Lothar Schäfer does, "What quantum physics reveals about how we should live." It will be scandalous in many quarters that he makes such a statement. Physics is proud *not* to apply to everyday life. You and I raise a family, pursue our careers, and face our everyday challenges without the slightest need to refer to the equations of Isaac Newton and James Clerk Maxwell. Every cell in the body obeys the principles of physics discovered by these pioneers and others, but the fact that physics speaks in mathematics removes it from the experience of living.

Ironically, the gap widened enormously with the arrival of quantum physics over a century ago. On one side, quantum mechanics revolutionized the concept of the physical universe, turning certainty into probabilities and unseating the very notion of solid particles. But on the other side, everyday life goes on. How can Nature contain two opposite worlds?

The everyday world, which is usually labeled Newtonian, surrounds us through physical objects; our five senses feel, see, and hear what goes on in the Newtonian world. Looking out into space, the vast distances between the stars may awe us, and the empty void may send a shiver through us. But planet Earth is the offspring of a familiar cosmos, in that even the most distant galaxy is a secure physical *thing*.

The other world, the quantum domain, couldn't be more different. Our five senses cannot detect it. The existence of Nature's building blocks begins in virtual reality, that is, in a realm of reality that isn't accessible to our senses. All things are rooted in the unreachable realm of no-thing. That the entire cosmos could arise from a place that is beyond human experience and unknowable, fits amazingly well with a saying from the Vedic scriptures of ancient India: "It is that which is inconceivable but by which all things are conceived."

The familiar and the inconceivable. That defines the paradox of modern physics in a word. It is sometimes called the mismatch between the microscopic and the macroscopic, meaning that what holds true at the tiniest scale in Nature, where the visible world emerges out of a realm of invisible forms, does not hold true in our world of plants, animals, rocks, and human beings. Worse, the quantum world is more fundamental. It cannot be dismissed as a queer anomaly. Creation is born here; genesis is now, taking place every nanosecond, with the inconceivable void as our womb.

Even though the conscious mind cannot cross over the border where matter ceases to exist (it is the quantum equivalent of Hamlet's "undiscover'd country from whose bourne no traveller returns"), Schäfer argues that we are most at home there. The fact that we are conscious beings derives from the womb of creation, the void that contains nothing we can describe in words. But we can't describe ourselves in words, either. Not when it comes to consciousness. Consciousness simply *is*. Each person is aware of being aware. In that regard, Hamlet asked the wrong question. It's not "To be or not to be?" but "To be conscious or not to be conscious?" As hard as it is to imagine death, it's infinitely harder to imagine having no mind.

The mind is maker of reality. It doesn't look on Nature like a customer looking in a department store mirror, intrigued by a reflection. You and your reality are one. If a sunset looks beautiful in your eyes, nothing in physics or biology can explain why. Certain wavelengths of light strike the retina of your eye. Electrical and chemical signals travel along the optic nerve, reaching the visual cortex in the occipital lobe. Going through several stages of neural processing, the photons that began as emissions from the sun become a sunset.

What's astonishing—and a complete mystery—is how you see

the sunset as a picture in reality, for there is no light in the brain, no images of a sunset. The brain isn't a camera, not remotely. Everything you see, hear, touch, taste, and smell must be created by the mind. Other living species don't confront the same reality as you. A hummingbird or honeybee, because it possesses an entirely different nervous system, lives in an inconceivable reality. Even mammals closer to us on the evolutionary chain inhabit a reality we cannot penetrate. A porpoise's auditory center occupies 75 percent of its brain. A porpoise interprets reality through rapid clicking noises that are six times higher than human hearing, emitted at a sound level of 170 decibels (the same range as a space rocket on takeoff), and are spaced by a gap of 60 one-thousandths of a second. By comparison, when an Olympic swimmer wins a race by a few hundredths of a second, to a porpoise's ear that is as slow as a tortoise. But even more radical, if we used our hearing to navigate through the world, a backseat driver would say, "You almost sideswiped that truck. Are you deaf?"

Consciousness takes as many shapes as there are nervous systems to match. But this book goes much further than a description of mental processing. It argues that we don't see because we have eyes or hear because we have ears. The reverse is true. The mind created sense organs to explore the universe. As someone neatly put it, we aren't a machine that learned to think; we are thoughts that learned to create a machine. This total reversal feels so strange at first that it can lead to queasiness. Modern science is materialistic. It approaches the world as if the brain came first, like a super computer set up to detect the physical world. Things always come first if you are a materialist (or a physicalist, to use the updated term). The problem is that quantum physics proved beyond a doubt, over three generations ago, that things didn't come first. First came the void (the pre-created universe), then the primal universe when the Big Bang occurred, and the primal universe must be divided into its first, highly chaotic instant before the laws of nature had coalesced (lasting, as some contemporary estimates suggest, less than 10^{-43} seconds), which bridged immediately into the unbelievably organized state where gravity and other laws begin to construct a coherent cosmos out of energy and then matter. The sequence is literally faster than light, or more correctly, faster than time.

Yet *something* organized the sequence from void to chaos to a coherent reality. The standard answer for that *something* is pure, random chance. Schäfer is a tactful, at times gentle, writer about the standard model of quantum physics, but random chance makes absolutely no sense as the organizing agent behind creation. To accept it requires a leap of faith. Since it is clear that the universe has evolved in such a way that chaos turned into order, there must be an explanation of how this occurred. The constants that mesh to keep the physical world from collapsing in on itself or, at the other extreme, flying off in all directions, are tuned to thousandths and sometimes billionths of measurable degrees. Randomness can't explain how such astonishing mathematical harmony came about; still less is it the best explanation. By analogy, is it plausible that you could take two gallons of sea water, fifty pounds of chemical solids (chiefly carbon), and assorted trace elements, throw them into a blender, and expect to pull a human adult out of it in the end?

The counterargument is to begin at the end rather than the beginning. Start with what we know we are, creatures with life, intelligence, creativity, orderliness, and evolution being expressed at every moment. If those qualities already existed as potentials in the pre-created state of the cosmos, the story of its unfolding is far more believable. The swirling chaos of the primal universe can then be seen as a passing stage on the way to orderliness, the way that boiling soup, once eaten, turns into the orderliness of our bodies, or the smoosh of paint on an artist's easel turns into a portrait of King George III.

By taking the word *potential* and following it all the way from before the Big Bang to the present moment (in the dimension of time), from the void to the expanded universe (in the dimension of space), and from the possibility of thought to the rich panoply of human desires, hopes, ambitions, and designs (in the dimension of mind), Schäfer delivers a conscious creation. Such a creation is more our home, as human beings, than the mathematical model of modern science—it embraces experience instead of data. In 1930 a famous conversation occurred between Einstein and the famous Indian writer and seer Rabindranath Tagore. The world press was literally camped on the porch of a tiny cottage in eastern Germany, awaiting what the two great minds would produce.

Tagore had won the Nobel Prize in literature in 1913, Einstein the Physics Prize in 1921, but beyond their specific fields, they represented one conception of reality—mystical, spiritual, subjective—and its opposite—scientific, materialist, objective. Each tried to convince the other of the rightness of his worldview. Tagore argued that we live in a "human universe" whose reality depends on us, the beings who inhabit its center as an expression of cosmic consciousness. Einstein, who was respectful and astute about such a view, maintained that the existence of a physical universe that preceded human beings was his "religion." He chose the word carefully, because he knew that scientific data don't describe reality; they describe only measurements, fragmentary snapshots expressed in the language of mathematics.

Tagore is no longer a household name (although he should be), while Einstein and the scientific worldview triumphed in every aspect of life. You would think that the question "What is reality?" has long ago been settled. But this book, along with a handful of others written by far-seeing physicists and spiritual thinkers, proves that we are far from knowing how reality works. To get any closer requires a personal journey into the domain of consciousness that is our true source. The really outrageous word in the book's subtitle ("What quantum physics reveals about how we should live") is "should." Schäfer doesn't mean this in a moral, lecturing way.

Instead, the way we should live is according to our highest potential. What this book holds out is a conscious universe where human potential is infinite, because the evolution of mind never stops, and our true nature is mind itself. To be frank, the human potential movement, which has existed for many years, rests on a mushy foundation. In a scientific age, no amalgam of scripture, gurus, saints, and enlightened souls will ever be adequate to transform people's lives, not on a mass scale. Schäfer's scientific approach has a far better chance, because his voice is the voice of consciousness itself. He provokes self-awareness in the reader by unfolding his own self-awareness, examining it from its very roots, celebrating where it might lead him. Many books offer themselves as invitations to a journey. This one is rare, because its invitation comes from the heart of creation, where consciousness has no audience except itself, and the end of the journey is to comprehend eternity.

THE PHYSICS OF ENLIGHTENMENT

You contain infinite potential and a boundless capacity for happiness and fulfillment. How do I know? Quantum physics has revealed this to be the nature of reality, and in this book, I hope to explain these discoveries and their profound implications to you. These principles can guide you to a better life and all of us to a better world.

The first step on the road to happiness and fulfillment is to understand the order of the universe and your place in it. It takes two to be happy in this world: It takes you and the world. And the two must be in harmony. You can't hope to find happiness in your personal life if you aren't living in harmony with the universe, and that simply cannot happen if you don't understand the universe you live in. This is why understanding quantum physics is so essential; it will help you understand the universe.

Most of my career has not focused on things such as happiness and fulfillment. I have spent the largest part of my life in a laboratory, doing the things that a decent physical chemist should do, and teaching. My focus in those years was on esoteric things such as quantum-chemical computations and electron diffraction studies of molecular structures. But then, a couple of decades ago, it struck me as a real problem that my students in physical chemistry had great difficulties understanding the conceptual meaning of quantum chemistry. They had no problems learning and applying challenging mathematical techniques, but they had no idea what it all meant.

I realized this was a common problem. Most people don't understand quantum physics even though we interact with it every day. For example, chances are that you have no problem using high-tech toys such as cell phones and computers, but chances are that nobody has ever tried to explain to you the principles of quantum physics, which enable these devices to function. To help my students understand what the concepts of quantum theory say about the world and the important life lessons we can learn from them, I developed a special "Quantum Reality" class. And with this, something unexpected happened.

We have, at our university, a course evaluation system. It provides every teacher with the anonymous comments of the students in a class. Students love this opportunity to give their teachers a hard time and to speak their mind without any fear. Now, in the Quantum Reality class, to my delight, students' reactions were overwhelmingly positive. Year after year they expressed their excitement about the power of the quantum phenomena to open their minds to a new understanding of the world and to change their outlook on life from what they had been taught before. "This opened my eyes; this has changed my perception of everything; this made me think of the world in a way I never thought about." These are just some of hundreds of similar reactions.

Through my students' comments I became aware of the life-changing power of the quantum phenomena. There is more to quantum physics than physics. I realized that understanding the quantum phenomena was useful for everybody—including me. They have changed my own life in many unexpected ways. For example, when I grew up in Europe, I was turned off by the dogmatic and threatening posture of the church to which my parents sent me, and I became a thoughtless atheist. Since then, the quantum phenomena have taught me that atheism is a mistake because it isn't in agreement with the nature of the world. Similarly, some forty years ago, when I came to the United States as an immigrant, I was amazed by the level of aggression that is accepted as normal in our society. As a child I had been constantly admonished to be kind and cooperative, yet in this country aggression is a virtue—certainly in the business world, in politics, and in public entertainment. In contrast, the nature of the quantum world shows us that the best way to live is with kindness, and knowing this has helped me to live in an aggressive environment without being aggressive myself. The quantum phenomena are so life changing because they reveal the connection between our life and the life of the universe.

Many people react with panic when they hear the words *physics* or *chemistry*. Everyone seems to have some horrible memories of a sadistic science teacher who terrified students. Shame on him or her! There is nothing to fear from chemistry or physics, and from more than forty years of teaching, I know that anyone with an interest can develop a sound understanding of the concepts of quantum science. It is true that the mathematical techniques involved are forbidding, and special training is needed to understand them. But you don't need to know the mathematics to grasp the concepts of quantum theory. They are challenging but in a different way from what you would expect.

The quantum concepts are challenging because what they say about the world isn't what you see on the visible surface. They seem so esoteric because they refer to an invisible part of reality that is real, even though you can't see it. When you hear, for example, that the quantum phenomena tell us that all things and people are interconnected in a holistic background of the universe, whose nature is mindlike, your first reaction might well be "That doesn't make sense to me!" Why doesn't it make sense? Because when you take a look at the world, you see isolated material things—lumps of matter—but nothing mindlike and interconnected. So your judgment is simple: "The quantum view of the world doesn't make sense!"

In this book we will often consider the hidden meaning of words, and how language can make you think in a certain way. Here is a wonderful example. Consider the statement we just used: "That doesn't make sense." What is this sentence trying to tell you? It is telling you that, if it isn't in your senses, it is meaningless; it is nonsense. But why should being in your senses be synonymous with being meaningful? Why should something be meaningful only if you can sense it? And here is the thing: The phenomena of quantum physics tell us that the most important part of the world is invisible; that is, outside of the realm of your senses. This part transcends our experience, and it can never be in your senses. It means that, in a very fundamental way, the invisible realm of reality doesn't make sense. But the invisible part of reality is real, because it can act on you. For example, it can make the neurons in your brain jump from one quantum state to another and give you either a brilliant idea or a headache! So not being in our senses—not making sense—doesn't mean something is meaningless.

Another reason why the quantum phenomena may seem so challenging to understand is that their discovery is relatively recent. Quantum physics began in 1900, but it took decades to evolve to its fullest, and important aspects are still being discovered. Quantum physics is a branch of science that deals with the elementary constituents of things at the foundation of the visible world, such as atoms and molecules. At that level, surprising aspects of the world are found: Energy, for example, appears in indivisible amounts, called *quanta*. That term is the plural form of the Latin *quantum,* which means "how great" or "how much."

Because of their recent discovery the messages of the new physics have never really been accepted in the public domain. Journalists don't want to write about things they don't understand, so they don't write about the quantum world. Teachers don't like to talk about things they don't really understand, so the science that is being taught in our high schools is still mainly Newton's physics of materialism and Darwin's biology of selfishness and aggression. Chances are that, when you grew up, you were taught these theories and their corresponding view of the world. But quantum physics has now shown these to be in fundamental ways inaccurate and even wrong. The early years of your life were formative years. Everything that you heard when you were growing up was downloaded into the depths of your mind, where it now influences—even controls—your thinking. So, when you are confronted with an aspect of the world that doesn't fit those theories, the first reaction of your mind may be to reject it, and a good way for it to do that is to tell you "You can't understand this." It isn't that the concepts of quantum theory are in themselves so difficult. No, the problem is that your mind makes them difficult.

You can think of the biases in your mind like a bacterial infection. In the same way in which your guts at one time were invaded by E. coli, your mind at one time was infected by a population of bacteria. These roam the rivers of your mind like predators: Every idea or thought that they don't want you to think is snapped into pieces. The goal of these predators is to close your mind to new ideas. You can see this in public discussions, when people argue on different sides of a controversial issue. In such discussions the focus of the participants is never on learning something new, but on protecting their personal bias. Max Planck, the founder of quantum physics, has described this phenomenon. "A new scientific truth," he wrote, "does normally not prevail in the way that its opponents become convinced and declare that they have learned something, but rather because its opponents eventually die out, and the following generation is familiar with the truth from the outset."

If your mind is controlled by a bunch of predators that force you to live in conflict with the order of the world, it can make you very unhappy: It can even make you physically sick. This book is about making a new beginning.

Making a new beginning means digging deep down into your mind to get everything out that is in there—your wishes, your convictions, your views of the world, your motivations, your principles, your logic, your habits, your love and needs, everything—and ask yourself: Why is all this stuff in my mind, and why do I want it to be there? Did I put it there, or has someone turned me into a puppet?

This book will give you options, allowing you to make an educated decision as to what to accept in your mind and what not; and lessons on how to keep an open mind. When we talk about the power of the quantum phenomena to be a guide in our life, even in moral and spiritual issues, one aspect is important. The lessons from quantum reality will never tell you what you have to think and what you have to do. The lessons are never dogmatic like the rules of a catechism or biblical commandments; they provide you with options—with suggestions of how you can live. When I taught my Quantum Reality class, I always emphasized to my students that I would never tell them what they should think. But I would describe options to them that would make them think: alternatives that they could consider in their private decisions of how they wanted to live.

Making a new beginning is something like the awakening that René Descartes, the French philosopher often called the founder of modern philosophy, experienced in the seventeenth century. One night, when he was a soldier with the French army in Germany, he saw in a flash the order of the universe and "the multitude of errors that I had accepted as true in my earliest years, and the dubiousness of the whole superstructure I had since then reared on them." He observed "the consequent need of making a clean sweep for once in my life, and beginning again from the very foundations." Once in a lifetime, or maybe several times, each one of us, you and

I, should have a Cartesian moment in which we expose the errors that we have accepted as true since our earliest life and plunge into the "deep sea" of our mind.

Our minds need a clean sweep, and so does the globe. This world can't continue to live in the way in which it has lived for the past few hundred years. The worldview of the classical sciences—Newton's physics and Darwin's biology—and the attitude toward life that followed from it have ridden this world into a ditch. Like our individual minds, the whole world has to make a new beginning.

So let us take the first step. Quantum theory will open your mind to exciting possibilities. It will show you that you have a potential in you that is boundless, and it will reveal the connection between the physical and the mental, between the rational and the spiritual, between physics and metaphysics, between the quantum world and your morals, and between the order of the universe and the order of the human world.

When I first got involved with the quantum phenomena I thought that the discovery of quantum physics in 1900 was merely the beginning of a change of the worldview of the physical sciences. But everything changed at that time—not only physics, but also the arts, literature, music, the order of the world. There are now reasons to believe that the change that began in 1900 wasn't only a change in the worldview of science, but a mutation of the human consciousness. Why don't you take a jump—a quantum jump, of course—and join the new species?

Introduction

Your Cosmic Potential: Being Part of the Universe

There are many big questions to be curious about, and one of the most fascinating is this: What makes us human? I would say that everyone's life—yours and mine, Leonardo da Vinci's and Albert Einstein's, a first-time mother and her newborn baby's—is guided by an inner potential, and our supreme good lies in actualizing our potential. More than love and intelligence, or consciousness and joy of life, this potential and our deep need to actualize it are what makes us human.

Certainly, we take pride in our ability to love and the power of our consciousness, but they aren't specifically human—many animals are capable of love. It may not be hard science, but when you come home after a long day's work and your dog is jumping up and down around you and greets you with welcoming yelps, you know that this animal loves you. Many other creatures besides us are conscious of the world. A lion on the prowl is fully aware of what it is doing, and so is the frightened gazelle in its sights. We feel unique in possessing intelligence—at his gloomiest Hamlet can still look at the human race and exclaim, "How noble in reason, how infinite in faculties." But, in reality, intelligence isn't specifically human, either—even plants need a certain level of intelligence to live. And as to joy in life, who would deny it in a colt dancing and prancing across a sunny meadow on a summer morning? No, what makes us human above anything else is this mysterious potentiality in us.

Everything we have built in the material world springs from that hidden urge, our inner sway and creative source. Our inner potential makes us who we are.

You can go far beyond your conception of who you are and what you can achieve. It's not as if human potential is a new subject; there is a whole movement built around it. But for most people, when it comes to achieving their full potential, they have barely begun to scratch the surface. In this book I will show that the potential in you is part of a cosmic potentiality that exists in every particle of the universe. It exists as a field of states of what is possible in the world, and it is as real as gravity. Gravity is invisible, yet potentiality is even more hidden from view because it exists at a level that you could call the *mind level,* and science will never be able to measure it with instruments or take photos of its landscape. There is no direct experience of it, because its states are possibilities that are waiting to appear in the visible world. Yet, there is enormous power in it. Everything that exists in the visible world has first existed as a state in the cosmic field of potentiality. Nothing comes out of the blue; everything emerges out of the cosmic potentiality. We build our dreams, hopes, and visions on what is possible: finding perfect love, ending war and violence, feeling the presence of God. What would be important to learn, if it can be done, is how to use our mind to tap into the cosmic field of possibilities, in order to make our dreams a reality.

THE POWER OF MATHEMATICAL FORMS

I have used the mind level to characterize the cosmic potentiality. That description will help us unfold some exciting aspects of what the cosmic potentiality is and how it relates to our own potential. I define *mind* as the realm of inner images in us that guide our life. In our mind these images find a special level of reality in our consciousness. The concept of the "inner images" in our mind derives from psychology and was introduced by brain researcher Gerald Hüther to describe "all that, which is hidden behind the external, visible and measurable living phenomena and guides and steers the reactions and actions of living beings." There are images in you that help you to be a good citizen or a caring mother, or guide you in moving around in the world. Your inner images determine what you think of yourself and what you think of the world and of other people, and so they determine how you interact with the world. Everything that you can do, every step that you can take, is guided by an inner image. Since you can undertake nothing that isn't first initiated and then allowed by an inner image, these images determine what is possible for you: They define your potential.

The inner images arise in your mind from a variety of sources. Some act through your consciousness, while others are hidden in your unconscious. Some of them are inborn. Others arise in activation patterns of neural networks that your personal experiences have created in your brain. And yet another group appears in your mind out of a nonpersonal and collective realm of forms with which we are all connected, as Swiss psychiatrist Carl Gustav Jung described in the last century. Whatever the source of the inner images in us may be, the fact is that behind the physiological processes of our brain, nonempirical forms are at work that control our actions and give meaning to our life. These forms aren't visible to or measurable by brain scientists. They are true forms of potentiality, because they exist before they manifest themselves in the visible world by affecting your body and controlling its processes.

You can think about the states of cosmic potentiality in the same way you think about the inner images of your mind. The cosmic potentiality is a field comparable to other fields spread throughout the universe, such as the gravitational field, except that cosmic potentiality doesn't contain any matter or energy, just states of what is possible in the world. What exactly do I mean by the *states of what is possible in the world*? We have to dig in deeper to get a clearer understanding, but it is worth it: Getting a grasp of potentiality is one of the most important things you can do for yourself in this life.

To appreciate the answer that physics has given this question, we have to keep in mind that the states of potentiality are invisible, and nobody can give a definitive description of invisible things. Physics can only observe what comes out of these states and then work backward, extrapolating into the darkness of the unknown, and *suggest* what its states *might be like*. So far, the best that physics has come up with is the suggestion that the states of the cosmic potentiality are *forms,* meaning *mathematical forms* or numerical patterns: not patterns of energy or matter, but just plain *numbers*. For the lack of anything better, we will accept such descriptions, but we must always keep in mind that we can't really know what the nonempirical realm of the world is like. It can interact with our world in space and time, but we can't even know that space and time apply to it, inside it. Perhaps the best way to think of the nonempirical realm of the world is like how many mystics have described their experience of God: Their experience wasn't of anything that appeared to their senses—it wasn't anything that they could feel, hear, or see—and yet it was real and powerful.

It has often been said that the American legal system is a formalistic system in which the notion of justice has been lost. Right or wrong don't matter, true or false don't matter; the only thing that matters is to follow the letter—or form—of the law, and to do that in a formalistic way. You can think of quantum physics in the same way: It wasn't invented as an *intuitive* or *visual* description of the world, like you might expect, but as a *formalistic* description, and, even though nobody wanted that, in the process of developing this formalism, the notion of reality was lost. The first concern of the physicists was to find a formalism that fit the visible phenomena so that we could deal with them. Once that job was done, they could deal with the problem of what it all might mean for our understanding of the nature of reality.

Interestingly, the physicists of the twentieth century weren't the first to suggest that the world consists of numbers. Already in the sixth century BCE, Pythagoras, one of the Greek founders of Western philosophy, claimed to have found irrefutable arguments for the thesis that "all things are number," meaning that the essence of things isn't found in some stuff, but in forms. And in the fourth century BCE, Plato, one of the most influential philosophers of all times, taught his students that atoms were *mathematical forms*. Contemporary physicists are, of course, not guided in any way by the arguments of historic philosophers, but they find that they need something such as the Pythagorean formalism or Platonic idealism to explain the visible world.

The English word *image* is usually understood to mean an image *of* something, presenting a likeness with an object or structure that it images. In a general way, the word *image* can also be understood as a graphical form or pattern not *of* something, but a formal motif or figure in its own right. An abstract painting in modern art, for example, can be such an image that an artist put down on canvas without having any specific thing in mind. In this sense, the states of the cosmic potentiality can be described as images: They are the inner images of the universe that guide its visible phenomena.

The images in the cosmic potentiality are, of course, different from the images in our mind. Nevertheless, you will not go very far from truth if you think that the cosmic potentiality is something like a mental domain of the universe whose inner images are found everywhere in the world. Human beings have a body and a mind because the universe has a body and a mind, so to speak, even though we never use such terms in connection with the universe. The intent of this choice of words isn't to revive archaic descriptions of the universe, but to demystify the human mind: There is nothing special about its images. Its images are as physically real as the images or mathematical forms that exist in atoms and molecules and determine their physical and chemical possibilities. Chemical reactions depend on the inner forms of the molecules that participate in them. The inner forms of atoms determine what kind of molecules can form. The visible world is shot through with invisible images that define its future possibilities. The aspect that is so difficult to get used to is the power of the images: *All* visible things and phenomena *need* an underlying image, or they couldn't have appeared in the visible world. Since this rule applies to the processes of the physical world as well as to the processes of our mind, it reveals an unexpected congruence of the physical and the mental that is typical for the new physics. In a metaphorical way, you could say that quantum physics is the psychology of the universe.

Don't Let Your Language Wash Your Brain

At this point you might object: "You want people to believe that underlying the visible phenomena of the universe there is a realm of the physical reality that doesn't consist of material things or fields of energy, but of ideas or mathematical forms, and that these forms are real, but nobody can see them? No matter, no energy, just ideas, mathematical forms, or patterns that exist as hidden states in all things, and everything that we can see comes out of them? You must be kidding me!"

Your reaction, if you indeed reacted in this way, is quite normal. There is something in us that leads us to expect, without even thinking about it, that only material things are real; that things, to be real, must be made of stuff—matter—to be of significance. I think that our language has a lot to do with this attitude, because it tells us that if a thing isn't important, it *does not matter*. What does that mean? It means that if it isn't made of matter it isn't important.

Think about the way you express yourself. What do you mean, for example, when you say something *doesn't matter,* or that it is *immaterial?* You are effectively saying that the thing to which you are referring doesn't contain any matter; but that isn't what you mean. You mean that it isn't important. So somehow your language leads you to think that to be material and to be important are one and the same thing, but that isn't necessarily true. Think of how many things are important in your life that have nothing to do with matter. To some extent, all languages have a built-in hidden ideology, washing people's brains in a way that they don't notice. Materialism—that is, the belief that only matter is real—is a brainwash of our language that tells us that only materialism is a reasonable way to look at the world. Thus, it is possible to think that Newton's physics and Darwin's biology, two of the most powerful formulations of materialism of our history, weren't inspired by the facts of nature, but by the language of their authors. So if you find it annoying that some esoteric branch of physics—quantum physics—wants you to believe that images or mathematical forms aren't irrelevant, even though they are without substance, then consider how your language influences your attitude.

There is a general lesson to be learned from this. When people disagree on some issue and try to find a consensus they will often propose to consider a given matter "objectively." But we can be objective about anything only to the extent that our language will allow us to do so. In a globalizing world this is a Babylonian confusion and a challenge that we have to deal with.

Reality Can't Be Denied

If the phenomena of quantum physics—the nonmaterial forms, the invisible reality, and so on—disturb you, you are in good company. Even the pioneers of quantum physics, all of them great minds, often reacted with some sort of irritation or frustration against the outrageous principles they discovered. More often than not they were overwhelmed by their own findings, and they tried to deny their implications or explain them away. But that is the thing about reality: It is what it is and it can't be denied.

Niels Bohr, for example, the twentieth-century Danish physicist to whom we owe the understanding of the structure of atoms, brushed aside the strange aspects of the quantum world by claiming that they had nothing to say about what reality really is. It is wrong to believe, or so he argued, that our experience of the world can tell us what the world is like. If you wanted to put his argument in a nutshell, you might say that you have no experience of things, but only of your experience of things. Thus, watching the world can tell you nothing about what the world is like. If your observations don't make any sense, just forget about them.

If you think about it, this view isn't so stupid. Crying foul when it comes to the correspondence between the experience of things and the nature of things makes a lot of sense. The objects around you, for example, are not like the light waves that bounce off of them and hit your eyes and make you see them. Similarly, the sensation of heat is not of the nature of heat. In physics, heat is motion. The atoms in things are constantly in motion. When a thing is hot, its atoms move faster than when it is cold. When you put a pot of water on a hot plate, you accelerate its molecules. Nevertheless, nobody has offered his or her mother-in-law a cup of freshly brewed tea and heard her scream out, "Ouch, too fast," when she burned her mouth. The experience of hot and cold isn't of fast and slow. And so it goes, Niels Bohr said, with all experience: It tells us nothing about the essence of things.

Albert Einstein, one of the greatest physicists of all times, wasn't a friend of quantum physics, either. He had a kind of reasonable, commonsense view of the world that didn't mesh well with the quantum theory. Throughout his life Einstein claimed that there was something wrong with this theory. Since he couldn't deny its technical successes and its precision in predicting the outcome of measurements, he accused it of being incomplete; something, he said, was missing. Once the missing links, or "hidden variables," were found, reality would be "reasonable" again. Einstein was an inspired man and not often off the mark. As it turns out, something is, indeed, missing in the world of atoms and molecules. However, whatever it is that is missing isn't missing in the theory; it is missing in the visible reality. The problem isn't that the theory is incomplete. The problem is that the visible surface of things is incomplete because it has little to say about the nonempirical realm of reality, where the cosmic potentiality has its home.

Einstein's protests, Bohr's skepticism, and the qualms and raised eyebrows of many other pioneers couldn't fix the formalism of quantum theory or stop its triumphs; the harder they tried, the more they discovered outrageous aspects of the physical reality. The specific forms that quantum physics had to adopt to describe the states of potentiality are *waveforms*. In the following chapters I will describe why this choice was necessary. This formalism implies that the things we see in the world aren't made up of material particles, but of waves; and that the universe is an ocean of waves—not waves of matter or energy, but nonmaterial, invisible waves in the realm of potentiality. There are indications that these waves are hanging together like the water waves in an ocean, so that the nature of the cosmic potentiality is that of an indivisible wholeness—some call it *the One*—in which all things and people are interconnected. The things that you see in the world are somehow actualizations of waves; they are emanations out of the One.

We mustn't forget that nobody can tell what is going on in the realm of potentiality and what its forms are. When we say that we can think that the realm of potentiality consists of waves, we mean that its interactions with the visible world show the properties of waves.

You Are a Crest on the Surface of the Ocean of Potentiality

It seems that we are proceeding here from one scandalous claim to another. First we are being asked to believe that our personal potential is part of a cosmic potentiality. Then we are asked to believe that the cosmic potentiality is a field of invisible states, like ghosts, but real ghosts to which we owe our existence. And now, we are supposed to believe that all the separate things and all people of this world are somehow connected in this invisible field: ordinary things such as the table I write on and the furniture in your home, your grandmother and a shepherd in Anatolia—all of us belong to the realm of the One and are interconnected in it. Give me a break!

The choice of words here is echoing what Einstein had to say about the empty forms and potentiality waves of quantum theory: Always ready to make fun of them, he called them *fields of ghosts*. This shows that he took them seriously, or he wouldn't have wasted any time on them.

Let's deal first with the possible interconnectedness of seemingly separate things. You can think about this in the following way. If you've ever flown over an ocean or large body of water, sometimes, when the weather is right, you can see some white patches on the water below: tiny white dots that suddenly appear, like out of nothing, move around, one this way, one that way, and fade out of existence again. It is really a beautiful sight and a fascinating play of forms, and it is easily understood: The white patches are the crests of waves in the agitated ocean. Now, here is the interesting thing: You can't see these waves, because you are too far up; you can see only the crests. But you have no doubt that waves are there—you may even say that you *know* it—and if anybody tried to tell you that the white spots on the water were some sort of big white fish, you would not believe it. Furthermore, when you see how each of the patches is going its own way and couldn't care less about what the others are doing, you can't be fooled into thinking that the waves are isolated and independent appearances, because you just *know* that the waves in an ocean aren't independent but conjoined in one giant movement of the whole.

This is a useful experience, because now, when you look out into the world and see the things and people around you, you will not be fooled into thinking they are separate and independent appearances because you know that they are the patches on the surface of an agitated ocean: the ocean of potentiality. In this nonmaterial ocean, which seems alive and more mindlike than thinglike, the waves constantly build up to new empirical possibilities and perhaps even to new forms of thinking that may find consciousness in a human mind. After each transition from "the possible to the factual," as one of the great pioneers of quantum physics, Werner Heisenberg, described it, the evolution of "tendencies or possibilities" for future actual events starts anew, but now from a different starting point than before, so that the universe is evolving. Physicist Hans-Peter Dürr, longtime coworker of Heisenberg's, has called this a learning process. With the experiences that it makes through us, the cosmic potentiality is learning. It constantly manifests and reabsorbs. There is a continuous flux from the evolution of tendencies to their actualizations—empirical events—and from empirical events to new tendencies. Each new state of potentiality carries in it, like a stigma, the memory of the last state. You are a white crest on the surface of this ocean.

The Power of Nonmaterial Waves as Principles of Being

So here you have it in a nutshell: the worldview of quantum physics in its whole unexpected, breathtaking, and exciting beauty!

Philosophers use a lot of fancy words to express their thoughts. They do so because these words, usually derived from some concepts invented by the ancient Greeks, are short and to the point. *Ontology* is such a word. It denotes the branch of philosophy that deals, as the science of being, with the nature of being. *Ontos* in Greek means "being" or "real."

Epistemology is another such term. It denotes the part of philosophy that deals with the nature of knowledge. *Theology* deals with the nature of the divine. All these words contain the Greek word *logos.* When you try to translate this word, you will find that it has had a whole spectrum of meanings throughout its history, including such terms as *speech, word, story, meaning, thought, reason,* and even *world law,* to mention just a few. It is an incredibly rich word and is also the root of *logic,* or the science of thinking correctly. What I have described in the last few paragraphs is the ontology of quantum theory. "In the beginning was the Word, and the Word was with God, and the Word was God." This is how the Bible begins its account of the creation of the world. In the original Greek text of the Bible, the term *logos* is used as the principle of creation, which is translated as "Word" in the English version. This is precisely the ontology of quantum theory. In the beginning—that is, before it is a visible thing—everything is a logical state in the realm of potentiality: *logos.*

From your school days, you may have some memory of the structure of atoms, which is as follows: All atoms consist of a nucleus surrounded by electrons. The nucleus contains practically all of the mass of an atom, more than 99.9 percent, and all of its positive electric charge. The space surrounding the nucleus is taken up by the electrons of an atom, which carry all its negative charge. You may also remember that your science teacher talked to you about electrons as elementary particles. When someone speaks of particles, instantly the inner image of tiny balls pops up in your mind, like of little round bullets or pool balls. Indeed, not too long ago, science textbooks presented pictures of atoms in which the electrons are seen as tiny dots running around the nucleus in circles. Well, you can forget about such pictures, because they are completely misleading and wrong.

Schrödinger's wave mechanics is currently the only theory that quantum chemists can use to calculate the properties of molecules from first principles; that is, from the properties of their atoms. It was named after the Austrian physicist Erwin Schrödinger, who developed this version of quantum theory in the 1920s. In this theory the electrons in atoms and molecules aren't permanent material particles that zip around the nuclei like planets around the sun; they exist in wavelike states, in which they are *standing waves, wave functions,* numerical patterns, or mathematical forms.

If this seems amazing, it is even more amazing that the waves into which the electrons in atoms evolve are empty. By that I mean that they don't carry any mass or energy, but just information on numerical relations: In these wave states the electrons are no longer material particles but patterns of numbers. Yet, the power of these waves is absolute, because the visible order of our world is determined by the way in which they interact or interfere with one another. The interferences of the wave functions of atoms, for example, determine what kind of molecules they can form. The interferences of the wave functions of molecules determine the forces that exist between them; in the cells of your body, these forces keep you alive. The shocking truth is that, at the root of matter, at the level of atoms and molecules, the notion of matter is lost in a realm of nonmaterial forms; and actuality turns into potentiality. Hans-Peter Dürr proposes: "Reality reveals itself primarily as nothing but potentiality." Schrödinger's mechanics is the basis of powerful computational techniques that are used by quantum chemists to calculate the properties of molecules. These techniques are so precise that some of the early calculations of molecular structures, which were performed in my research group a few decades ago, led us to numerous discoveries of errors in published *experimental* work. In the case of proteins—the molecules that are the basis of the chemistry of life—we were able to calculate important trends of their structures more than a decade before these were found in experimental studies. Consider what a scandal that was. Science is about performing experiments and accepts nothing but experimental data as a basis of knowledge. And then, all of sudden, some computational procedure, which works on the assumption that the electrons in atoms are waves, wants to tell us that experimental data can be misleading. It was a disturbing development and a shock to many! As one of the researchers whose work was corrected put it to me: "You should never have been allowed to perform such ridiculous calculations!"

What Is "Really" Real?

In 1911, Ernest Rutherford, a New Zealander working in England, discovered that atoms have a nuclear structure. Each atom consists of a nucleus, which contains nearly all of the mass of an atom and the entire positive charge, while the negatively charged electrons are somehow arranged outside of the nucleus.

Rutherford's nuclear atom was a shock, because the space taken up by the nucleus is about ten thousand times smaller than the average space taken up by an atom, even though a nucleus contains nearly all of the mass of an atom. This means that atoms don't fill space solidly, as the ancient Greek philosophers thought, but they are mostly empty space. To find out what I mean, just knock your head against a wall of your office. Go ahead, don't worry, you are just knocking on empty space! Actually, that space isn't totally empty; it isn't a vacuum, a void, or nothing: It is *filled with potentiality.*

Sir Arthur Stanley Eddington, one of the celebrated astrophysicists of the twentieth century, gives a wonderful description of the shock caused by Rutherford's atom: "The revelation by modern physics of the void within the atom is more disturbing than the revelation by astronomy of the immense void of interstellar space." Looking for a descriptive example, he finds it in the solar system: "The atom is as porous as the solar system." And, turning to the human level, he adds: "If we eliminated all the unfilled space in a man's body and collected his protons and electrons into one mass, the man would be reduced to a speck just visible with a magnifying glass."

Looking for an example of our own: If you would pack together all the nuclei of all the atoms on this globe, they would fit into a sphere with a radius of a few tenths of a mile—comparable to the radius of a football stadium and its surrounding parking area.

It is interesting that many languages have several words for what we call *reality*. In the German language, for example, there are two words, *Wirklichkeit* and *Realität,* that both have the meaning of reality, but they aren't synonymous. Rather, they describe different modes of being real. The first is derived from the German verb *wirken* (to have an effect) and the second from the Latin word for things, *res*. All material things are, of course, real, but nonmaterial entities are also real if they can have an effect in the world. The entities of the realm of potentiality in nature are of that kind; they are not things but forms. Nevertheless they are real, because they have the potential to manifest themselves in the empirical world and to have an effect in it.

"The entire cosmos is harmony and number," Pythagoras proclaimed, meaning a harmony that rests, as in music, on the ratios of numbers. Think of the vast empire of electronic gadgets that have changed our life. The cell phones, the computers, the transistors and integrated circuits: Without Schrödinger's discovery that the electrons in atoms can form standing waves, none of these gadgets could have been developed. As always, when a theory is technically successful, it whets the appetite for more; it gives rise to expectations that it is more than a mere tool that can put our observations to order, as Niels Bohr described it. When a theory is technically successful, there is a temptation to think that it has something to say about the nature of things. That is exactly what happened to Schrödinger's wave mechanics. At first it was a wonderful tool that rationalized the properties of atoms and molecules. Then it became a model of reality that led to the view that the basis of the material world is nonmaterial and that reality appears to us in two domains: potentiality and actuality.

THE COSMIC NATURE OF YOUR INNER IMAGES

Religions of all ages have always insisted that the essential reality isn't found in the material world, but in some transcendent part of the universe. However, all that you have to do to experience an invisible world is to look inside you, where your feelings and the images of your mind are real. As a person outside of your mind, I can't see its images, but I believe you when you tell me they are there, because I have them, too. So they are in you, they must be real, they must be somewhere. Real things are supposed to be somewhere! The question is, where exactly are the inner images in you? Asked about this, most people will point to their brain. The problem is that, when we open the brain and look inside, we see no images. That is how Gerald Hüther defined the inner images to begin with: "they are hidden behind" the visible living phenomena. The visible phenomena are the activities of the neurons in your brain. But the images they produce aren't visible. So, no big deal: The same thing that happens with the images in your brain also happens with the inner images in atoms and molecules. When we take these things apart and look inside, we can't see any images.

But we have to assume that they exist, because the actions of atoms and molecules can't be understood without this assumption. This is a great example that shows how something can exist in a state that is real, and yet it is, actually, nowhere.

An important property of the inner images of molecules is their indistinguishability. *Indistinguishability* in this context means that the inner images of all molecules of the same type are exactly identical. The inner images in molecules belong, on the one hand, to a specific molecule. On the other hand, they are exactly the same in all molecules of the same type. The exact identity of different things should already serve as a warning that something fishy is going on here: You can't find two objects in your environment that are exactly alike. Just think of anything! Two cars of the same make and type may look identical, but one may have a scratch under its body, and the other one not. Two salt crystals of the exact same size and shape may look identical, but one may have a hole inside its crystal lattice, where a single sodium ion is missing, and nobody can see that. No two objects of our ordinary environment are exactly identical, but molecules of the same type *are* exactly alike. In a cup of water there are more than 10^{24} perfectly identical water molecules: that is, a million-trillion-trillion indistinguishable things that are exactly identical down to the minutest details of their hidden forms. The identity of the inner forms in different molecules must mean that their logical order is not the personal achievement of a given molecule, but part of a cosmic order in things; we can call it a constitutional property of the universe.

Now, here is a stunning parallel involving the images in the human mind: On the one hand the images in you belong to your brain; on the other hand there is, as Carl Jung has shown, an important group of inner images that are identical in the minds of all people. They appear in our mind out of what Jung called the *collective unconscious*. This leads us to a question: Is the logical order of the images in our mind also a constitutional property of the universe? On the one hand, your inner images belong to you; on the other hand, they are on loan from the cosmic potentiality.

With this we have reached a defining point of our humanity whose implications are giant. If the inner images in us are cosmic, it could mean that we are connected with the cosmic potentiality. It could mean that we are part of a cosmic field that is acting in us. Of course, spiritual teachers of all times have told us all along that something like this is going on: God is in you. But to discover hints of a cosmic presence and cosmic activities within us, in the context of science's description of the world, redefines the playing field. It allows us to consider the matter without archaic threats and darkness. If it could, indeed, be shown that our mind is an extension of the cosmic potentiality, this would open exciting vistas of our cosmic role and dignity that are in complete contrast to the ruling Darwinian picture of human animals. That picture wants to tell us that we are creatures of chance, diced together pieces of rubble, and meaningless structures in a meaningless world. But, as it now appears, we are not like that. We are singular points in the cosmic field of potentiality, which seems under pressure everywhere to actualize in the visible world; and in us it has found a special way to do so, defining a cosmic destiny for each one of us!

The Mindlike Aspects of the Cosmic Potentiality

Since it can interact with our mind, the cosmic potentiality must be mindlike. Does this mean that the cosmic realm of forms is something like the mind of the universe?

Many physicists react with anger to such questions. They have no place, or so these physicists argue, in public discussions of the meaning of science. However, whether you like it or not, it is an unexpected fact and characteristic of quantum physics that, during the past hundred years or so, a large number of physicists, among them distinguished pioneers, were inspired by their science to speak of the presence of mind in the universe, and they weren't afraid to conclude that consciousness is a cosmic property. This is a phenomenon that has no equivalent in the physics of the centuries prior to 1900, which is usually called *classical physics*. This phenomenon does in itself not prove anything, but it gives us a choice: We can shrug it off as a sign that the pioneers of quantum physics just didn't understand what they were dealing with—and I have heard a number of physicists express this view. Or we can open our minds and take a look at why these people said what they said, and perhaps we can learn something.

Physicists spend their entire lives studying the properties of matter. Take Hans-Peter Dürr, for example. "As a physicist," he writes, "I have spent fifty years—my entire life as a researcher—to ask, what it is that hides behind the material. And the result is simple: there is no matter! . . . Basically, there is only spirit!" If you think that Dürr went off the deep end, then what will you think of Sir Arthur Stanley Eddington when he writes "The universe is of the nature of 'a thought or sensation in a universal Mind.' . . . To put the conclusions crudely—the stuff of the world is mind-stuff"?

Eddington's colleague Sir James Hopwood Jeans takes us even further, into an astonishing idea for the time: "The universe begins to look more like a great thought than like a great machine." Remember, science is all about machines, instruments, and measurements. Setting up experiments, analyzing the findings in a rational way, and sticking to the visible world: That is the business of scientists, their training, their pride, and their life, and yet the concept of the universe as a great mind opens up vast new possibilities, for where do we belong in that great mind? Jeans perfectly understood the vistas he had opened up, for he goes on: "Mind no longer appears as an accidental intruder into the realm of matter."

We are the intruders that he is talking about. In the biblical sense, classical physics and its offspring, Darwin's biology, have all seen human beings as strangers in a strange land. The time has come to develop a more realistic view of ourselves and the nature of the world. It is now possible to think that, as I have described it before, "there is a covenant between our mind and the mind-like background" of the universe.

If we can make these ideas stick—that is, if I can describe to you some visible phenomena that support the claim that the stuff around us follows invisible patterns of potentiality that have a cosmic nature; and if I can produce some plausible arguments that we, too, follow patterns of potentiality that are derived from a cosmic field—then there will be enormous consequences for our understanding of who we are, what we are, and how we should live in this world. These consequences will turn upside down the views of the age of Newton's and Darwin's science, which have become endeared to the general public; and for those who are willing to accept the consequences, life will be richer, more meaningful, and, yes, happier.

Is Quantum Physics a Form of Spirituality?

When I asked at the beginning what makes us human, you may have heard echoes of William James's famous book on the life of religion, *The Varieties of Religious Experience,* which has had a profound impact since its publication in 1902. That echo was no accident but intended as a subtle hint about the spiritual background of the concepts used by quantum physicists to explain the world.

Physicists typically believe that their experiments have forced them to develop the concepts of quantum theory. Out of the experiments with elementary particles, atoms, and molecules, all these unexpected ideas arose of wave functions, quantum numbers, and so on. In reality, practically all of the unexpected concepts that quantum physicists are using to describe the world were invented by spiritual teachers thousands of years ago. The quantum numbers, the concept of potentiality, the principle of wholeness, the importance of waves as the source of the manifested world—all of these ideas have historically spiritual roots. Does that make quantum physics a kind of spirituality?

Religions of all times have warned us for thousands of years that there is a part of the world that we can't see but that is real nevertheless, because it can act on us. The idea that the basis of the visible world is a realm of potentiality that doesn't consist of material things but of mindlike forms can be found in the teaching of the Indian sages thousands of years ago and in Aristotle's philosophy. And the same is true for the concept of the wholeness of the universe, the assumption that the basis of the material world is an ocean of nonmaterial waves, and the claim that consciousness is a cosmic property. Practically all of the basic ideas of quantum theory have ancient spiritual roots and have been around for thousands of years.

Consider Pythagoras, the sixth-century-BCE Greek who surprised the world by declaring that things aren't made up of any stuff, but that *all things are numbers*. An important detail about Pythagoras and his followers is that the Pythagoreans were a religious sect. Their theory of numbers was connected with their spiritual teachings. I don't think that the quantum physicists have anything spiritual in mind when they consider that elementary particles are numbers, but it doesn't matter what they have in mind. The fact is that, by the way in which it describes the world, quantum physics has taken science right into the middle of historic traditions of spirituality.

Aristotle accepted from Pythagoras the view that form is an important principle of being. But he wasn't willing to completely disregard, as Pythagoras did, the importance of matter in the world. So, in his metaphysics, Aristotle developed the view that all things are mixtures of stuff and form. The table I write on has a certain form. It can have its form because it has the necessary stuff to express it. The flowers in your garden have specific forms, and they have the stuff to show them. In this way, Aristotle claimed, all visible things are necessarily mixtures of form and stuff. But he made one exception: There is one *pure form,* he said, that exists without any admixture of matter. That form is God. This part of Aristotle's philosophy is often called *matter-form philosophy.* "Being means to have been formed," Johannes Hirschberger, the prominent twentieth-century historian of philosophy, describes this aspect of Aristotle's philosophy, and "becoming means receiving form; fading out of being means losing form."

So pure forms are divine? But in Schrödinger's quantum theory the wave states of electrons in atoms are empty, nonmaterial waveforms; pure numbers. Does it mean that they are an expression of the divine?

Your thoughts exist in you, like the forms of the quantum states exist in molecules. Your thoughts are connected with electric activities in your brain, but the thoughts in themselves aren't electric energy and they aren't dancing material particles. Does it mean that they are an expression of the divine?

With your permission, I am just thinking out loud. These thoughts don't prove a thing, but they *suggest* a lot. Physicists don't have spiritual issues in mind when they design their experiments, and they shouldn't. At the same time, once they have put their findings before the world, the findings are out of their hands, and they shouldn't forbid anyone to draw their conclusions.

The actualization of things out of the cosmic potentiality can be described as an emanation out of a holistic background of the world; that is, out of an indivisible wholeness that is One. That concept isn't new, either, and has been used by philosophers for a long time. For example, it is important in the metaphysics of Plotinus, the third-century Greco-Egyptian philosopher who has become known for the way in which he revived Plato's thinking. Plotinus thought of God as "the One." In thinking about the origin of the world, he somehow developed the notion that God isn't the creator of the world, but the world *is an emanation* out of God, due to a necessary flowing over of the divine. As Hirschberger reports: "The One is all. All is out of the One."

Many other examples come to mind that show that a connection exists between contemporary physics and ancient spiritual teachings. I'd like to explore those connections together with you in this book.

IS QUANTUM PHYSICS A SORT OF IDEALISM?

Idealists in philosophy are people who believe that true being rests in an invisible realm of ideas and not in the visible world of things. In Western philosophy, such schools of thinking go back to Plato, who believed that all visible things are copies of ideas that exist somewhere in some invisible, transcendent part of the world. I am sure you will notice the similarity of Plato's views with quantum theory and its thesis that all material structures are actualizations of invisible forms. So, because it seeks the essential reality in an invisible part of the world, without any doubt, *quantum theory is a form of idealism*! It shares this aspect with Einstein's relativity theory, which also places the essential reality in an invisible space; that is, the four-dimensional space-time.

Augustine of Hippo also believed in the existence of a realm of immutable forms. He thought that such forms are needed to maintain the uninterrupted and reliable essence of things. He also believed that these forms don't exist somewhere on this globe or in the visible universe, but in the mind of God. If he were around today, he would probably think that, through the forms of the cosmic potentiality, divine thoughts are flowing into the human reality.

So let's face it: In the context of ancient spiritual teaching, the non-empirical reality is the liaison reality where the physical becomes spiritual, and the spiritual turns physical.

The concept of forms as principle of being wasn't a European invention; it's been used all over the world. For example, it is found in Buddhist philosophy. When Buddhists speak of *Alayavijnana,* they mean a storehouse in which the memories of all human beings—of their thoughts, feelings, wishes, and deeds—are stored as possibilities or seeds.

If everything that is empirical is an actualization of a form out of the cosmic potentiality then this principle must also apply to our consciousness and its contents. Thus, we can think that the cosmic potentiality is the source not only of the material things in this world but also of the principles of our mind. The common source, of external structures and internal principles, is the basis of our ability to understand the world.

One of the important abilities that our brain has acquired in the course of its evolution is its sensitivity to light waves. It has done that by developing eyes with which we can see. It is possible to think that the brain has also evolved some sensitivity to potentiality waves by evolving "eyes," or neuronal structures, that can receive signals out of the cosmic field and, in turn, take forms of our consciousness back into the cosmic field. In the first half of the twentieth century, Swiss psychiatrist Carl Jung developed a theory of the human mind that is based precisely on principles of this kind.

Jung described empirical evidence for the thesis that our mind can be affected by a field of invisible forms, which he called the *archetypes*. These forms can actualize spontaneously in our mind and influence "our imagination, perception, and thinking." He called them typical modes of apprehension, or psychic organs present in all of us. He thought that they motivate and guide our mind and give meaning to life. He called the realm where these forms exist the *collective unconscious*: "A psychic system of a collective, universal, and impersonal nature which is identical in all individuals. . . . It consists of pre-existent forms, the archetypes, which can only become conscious secondarily." Thus, the equivalent of the nonempirical realm of the physical reality is the unconscious in the reality of our mind, and the archetypes are a specific class of inner images of the kind that Gerald Hüther has described.

Like the inner images in molecules, Jung's archetypes are nonempirical entities because they "have never been in consciousness" before. In addition, the collective unconscious is a realm of wholeness. As Jung describes it, beyond the narrow confines of our personal psyche the collective unconscious is "a boundless expanse full of unprecedented uncertainty, with apparently no inside and no outside, no above and no below, no here and no there, no mine and no thine, no good and no bad." Jung was a scientist, but this is more than a scientific description of a "psychic organ." This is the account of an idealist and mystic describing his contact with the divine, a realm of wholeness, "where I am indivisibly this *and* that; where I experience the other in myself and the other-than-myself experiences me." All things are connected in this psychic realm, as they are connected in the physical realm of the cosmic potentiality: "There I am utterly one with the world, so much a part of it that I forget all too easily who I really am."

Take a second and think about this: The visible world around us exists because an underlying field of invisible forms defines the potential of the world. And now we find that you and I, too, can exist only because an underlying field of invisible forms defines our potential. Identical in all human beings at all times and in all places, these forms, archetypes, or ideas are universal: Why should it be outrageous to take this as a sign that the forms of our thinking also belong to some cosmic field? Is it such a long shot to think that Jung's realm of forms and the realm of forms of quantum physics are one and the same realm of the cosmic potentiality—a medium of spirit where our scientific, philosophical, and spiritual convictions are integrated in the nondual order of the One?

To many scientists such thoughts are upsetting. The view is widespread that science shouldn't get involved with such issues. It should be useful and technical, not inspiring; logical, but mindless. However, we should have the courage for an enlightened and liberated science that does more than serve stockholder equity. We must make an effort to understand the nature of all levels of physical reality: the empirical and nonempirical, the material and spiritual.

In the first few years of this century, France was rocked by an aggressive controversy involving the question of the evolution of life. I am sure that you are aware of such public fights in the United States between a religious public and a school of atheist scientists who love to stir the pot. Interestingly, the same conflicts in France are typically between an atheistic public and scientists who see more in the physical world than its visible surface has to offer. In this situation a group of scientists decided to go public with a "European manifesto." In 2006, it was published by *Le Monde,* one of the large French daily papers. "Religious or metaphysical ways of thinking," the manifesto begins, "should not, a priori, interfere in the ordinary practice of science. However, we also consider that it is legitimate, indeed necessary, to reflect, a posteriori, on the philosophical, ethical and metaphysical implications of scientific discoveries and theories."

In the eighteenth century the idealist philosopher Georg Wilhelm Friedrich Hegel developed the theory that the "Absolute" or "the self-motivated Spirit" is the basis of reality and everything that exists is an actualization of spirit. Hegel's philosophy is called *absolute idealism.* It got its name because spirit is the source of everything and creates everything; thinking and being, subject and object, the real and the ideal, the human and the divine—all are One. It opens up amazing perspectives: Your consciousness isn't your own, but the consciousness of the cosmic spirit; your thinking isn't your own, but the thinking of the cosmic spirit who is thinking in you; your potential isn't your own, but the cosmic potential to which you are connected. "Man knows of god only," Hegel writes in his *Phenomenology of Spirit,* "insofar as god knows of himself in man; this knowledge is god's self-consciousness."

At first sight it seems incredible that someone should come up with such ideas, but Hegel wasn't the first to express them. Thousands of years before him, the Indian sages invented the tale of the pots filled with water and placed in the sun. When the sun is shining on a million pots of water, it is in each one of them. But there is only one sun! In the same way, consciousness is in all of us, but there is only one consciousness. As Hegel describes it, "The spirit of human beings, to know of god is only the spirit of god himself." In addition, Hegel believed that God evolved with us in our history and in all cosmic processes of becoming. "The truth is the whole. The whole, however, is nothing but the essential being, which is perfecting itself in its evolution." If God's mind is in ours, it follows that his "words can be in our mouth," as the Bible describes it in Jeremiah 1:9.

If reality is an "undivided wholeness," as David Bohm, one of the pioneers of quantum physics, believed, everything that comes out of the wholeness belongs to it, including our consciousness. Basically, this is the argument that physicist Menas Kafatos and science historian Robert Nadeau used to support the view that the universe is conscious. From the perspective of contemporary psychology, British psychiatrist Brian Lancaster has summarized this view in the following way: "Consciousness amounts to a fundamental property, irreducible to other features of the universe such as energy or matter."

When we were born into this world, we were ejected out of the wholeness, and the experience was traumatic. The world of an embryo in the mother's womb is the archaic world of wholeness.

There is no inside and no outside: The growing baby and its surroundings are one. In the moment of its birth, when it is ejected out of the wholeness, the first thing that the newborn baby does is to let out a horrified cry. The biblical story of the eviction from paradise is a symbolic account of this existentialist crisis. But, contrary to popular accounts, paradise isn't a garden of sensual pleasures, but the invisible and mindlike realm of reality.

We have a need to be in touch with the wholeness, and there is a price to be paid when the need is neglected. "My experiences from psychosomatic therapy," writes psychotherapist and successful author Hanne Seemann, "have taught me that human beings who reside exclusively in the material and rational domain, will sooner or later develop psychosomatic irregularities, because their soul cannot bear this."

The longing for the wholeness is the source of our spiritual needs and the basis of all mystical experiences. Plotinus has given a touching account of such an experience: "Often when I wake up out of my body to myself and step out of the otherness into myself, I behold a most wonderful beauty. It is then that I believe in the strongest to belong to the greater destiny, and bring about with my force the perfect life, and have become One Thing with the Divine."

If the universe is One, all is out of the One, the One is in all, and the cosmic spirit is in ours. This is the message of the quantum phenomena, and it is Hegel's message: The cosmic spirit itself is thinking in us, becoming conscious of itself.

In the Gospel of Thomas we read in logion 117: "His disciples said to him: 'On what day will the Kingdom come?' And Jesus said: 'It will not come when you are looking outward for it. They will not say; "Behold it is there!" or; "Behold, it is that one!" Rather, the Kingdom of the Father is spread out upon the earth and men do not see it.'"

So, where is the kingdom? It is the nonempirical realm of the cosmic potentiality, and it is in you.

The Mutation of Our Consciousness

Now take a look at what happened to us in just a few pages of the introduction to this book: Starting with some ordinary and harmless concepts of quantum physics, all of a sudden we find ourselves in the thicket of spiritual issues. This surprising development is a shocking experience to many. It implies a congruence of the physical, the spiritual, and the mental that is difficult to accept for the average Western mind. In fact, its acceptance amounts to a change of mind that is so fundamental that it can be considered a mutation of the human consciousness, as though the evolution of life was taking a leap into a new human species.

Such mutations have happened before. In the last century, philosopher Jean Gebser has described how the human consciousness has gone through a number of mutations in our history, evolving from an archaic structure to the magic, mythical, and mental structures. At the present time, he believed, we are undergoing another mutation—that is, a change of our consciousness to an integrative structure. The integrative structure is characterized by the fact that it can reconcile seemingly incompatible views of the world, such as the spiritual and the rational, and the logical and the mystical.

Pierre Teilhard de Chardin, a French Jesuit priest and paleontologist who lived in the first half of the twentieth century, anticipated a similar process. "Like the meridians as they approach the poles," he wrote, "science, philosophy and religion are bound to converge as they draw nearer the whole."

Teilhard was an inspired man. Since his vision didn't agree with the official dogma of the church, he was forbidden to write in public about it. Persecution has always been the fate of innovators in Western history. Socrates was poisoned. Descartes had to flee from France and hide in Holland. Galileo was imprisoned in his house. Hundreds of thousands of people were burned at the stake to erase their ideas. The ruler of Prussia threatened to dismiss Kant from his teaching position.

Teilhard was sent into exile, first to China and then to New York. At the other end of the spectrum, the scientists of his time had, at best, a pitying smile for his theories. My heart reaches out to this man, who had to live his life in solitude and isolation. Suppressing a person's potential in this way is the most hideous crime. I think with sadness of his tragic life, and I admire his courage and the beauty of his insights.

Meister Eckhart was a medieval monk and mystic. He, too, was accused of heresy and put to trial, but a merciful God allowed him to die from stress, rather than fire, before the end of the trial. In what is now called his fifty-third sermon, he writes of God: "God is an unspoken word. God is a word that speaks itself." What I hope for this book is that the unspoken will speak to you through its words. If you think about it, *you* are an unspoken word, but a word that only *you* can speak. This is your potential.

With this, our task has been mapped out for us: to first understand the nature of reality as it is revealed to us through quantum physics and then to explore what that tells us about how we should live. Right now you may have more questions than answers, but in the following chapters, we'll unfold the relevant aspects of quantum physics in an easily understandable way. If this is the first time that you've heard about them, chances are that the wonders of the quantum world will change your view of life and give it a new meaning. They will certainly give you a new understanding of the infinite potential in you.

PART ONE

The

NATURE

of

REALITY

Chapter 1

MATERIALISM IS WRONG:
THE BASIS OF THE MATERIAL WORLD IS
NONMATERIAL

"Modern atomic theory is thus essentially different from that of antiquity in that it no longer allows any reinterpretation or elaboration to make it fit into a naive materialistic concept of the universe. For atoms are no longer material bodies in the proper sense of this word . . . the experiences of present-day physics show us that atoms do not exist as simple material objects."

—WERNER HEISENBERG

The phenomena of quantum physics force us to believe that the basis of the visible world doesn't rest on some material foundation, but on a realm of nonmaterial forms that have the properties of waves, as though our world were afloat on an invisible ocean.

Western philosophy was born in Greece in the years between 600 and 400 BCE. Basically, all the concepts and possible views of the world that have dominated the thinking of the Western mind originated at this time, when people were interested in finding some sort of *primeval stuff,* some primordial matter out of which everything else is made. It was the birth of materialism and of the concepts of elements and atoms; that is, the idea that all things are made up of some tiny units of matter. If you take a material object and divide it into smaller and smaller parts, or so the argument went, then you will eventually arrive at a level where you can divide no more, no matter how sharp your knife is. This is the level of the *indivisible* constituents of things: *Atomos* in Greek means "indivisible."

In a constantly shifting and confusing world, the Greeks searched for something lasting and trustworthy, and they believed they found it in stuff, matter. If stuff is the source and basis of everything, then it isn't amazing that the word *matter* has a connotation with *mother*. This connotation isn't found in its Sanskrit roots, but in Latin: Matter is *materia;* mother is *mater.* Matter is the mother of it all—something sacred—and materialism is its religion. Quite generally, words trigger inner images in you and affect what you are thinking. Their hidden meanings aren't accidental.

In the sixth century BCE, the city of Elea in Italy was an important center of learning. At that time one of its citizens, Parmenides, founded a school of philosophy whose teachings still affect us today. Parmenides added the concepts of space and time to matter. He asserted that stuff is eternal, indestructible, and unchangeable and fills space solid. This makes "being" and "nonbeing" the same as "full" and "empty." *To be* means *to fill space solid.* If something doesn't fill space, it isn't real.

These ideas dominated science for centuries. In his book on optics, Isaac Newton, for example, wrote about material particles that "God in the beginning formed Matter in solid, massy, hard, impenetrable, moveable Particles." This is exactly the point of view of the ancient Greeks. Newton took great pride in the fact that his science needed "no hypotheses" because he was dealing with facts—but it isn't clear how he knew for a fact the manner in which God at the beginning formed matter. He even went on to claim that the solid, hard, and impenetrable particles are "so very hard, as never to wear or break in pieces; no ordinary power being able to divide, what God himself made one in the first creation. . . . And therefore, that Nature may be lasting." With Newton, the doctrine of materialism entered the physical sciences and, after that, public life. Its connection with God's will confirms the impression that it has religious roots.

It is interesting that Newton spoke of elementary *particles* when he referred to the microscopic constituents of things. The roots of this word are related to the Latin *particula*, meaning a "small part" or "little piece," and to *partiri*, which means "to divide." You divide a thing; you end up with small parts: particles. It seems simple! The problem is that the concept implies that the particles that you find at the bottom of material things are as solid and permanent as the things that they form. But that isn't so. This is what we will have to work out in the rest of this chapter: The elementary particles at the bottom of things aren't lumps of matter in the ordinary sense of this word. As we shall see, they have wavelike properties, and the nature of these waves is closer to the nature of thoughts than things. So instead of calling them particles, it would be entirely justified to call these elementary building blocks *elementary waves,* or *wavelets.* And since these wavelets have thoughtlike properties, it would be perfectly all right to call them *elementary thoughts.* In this sense, your body is made up of elementary thoughts.

FROM MATERIAL PARTICLES TO WAVES

How many people do you know who can drive a car? Probably a lot of them! How many of them can drive a car safely, even though they know very little about how it works? Many people can drive a car even if they have no idea what a fuel injection system or a piston is. They may not even know about gears, if they first learned to drive on an automatic car. And yet they can drive safely from point A to point B. Something like this is what I propose for you to do in this chapter: move from point A to point B.

Of course, our focus in this chapter has nothing to do with

driving cars. Rather, the task we are facing is to take a piece of matter—some stuff—and turn it into numbers. At point A we are holding a material object in our hands—some massy thing— and then at point B this thing will have turned into a bunch of numbers. Mass gone! You can think that the numbers represent a mathematical form, such as a circle or a sphere.

Turning matter into nonmaterial numbers or mathematical forms is the easiest thing that you have ever done; you don't even need a driver's license. This is so because, at the level of elementary material particles, matter turns itself into numbers, spontaneously: All you have to do is to create the right environment and, bingo, it happens. If the conditions are right, elementary units of matter, such as electrons, atoms, and molecules, will spontaneously make transitions from a matterlike state into a numberlike state, in which they are no longer material particles but mathematical forms, patterns of information: very much like ideas.

What does this have to do with driving a car? Well, to get from point A to point B in the quantum world, you can proceed in the same way in which you drive a car: You have a couple of options. You can do it with the expertise of a mechanic, who understands all the technical details of the process that will get you from point A to point B; or, you can simply hit the road without worrying too much about the technicalities of driving.

To understand why the phenomena of quantum physics force us to think that material particles turn into mathematical forms when the conditions are right, you can proceed at different levels of technical insight: You can take some time out and get a PhD in physics, so that you'll understand what is going on at a level where you can take the system apart into its nuts and bolts and enjoy putting it back together again. Alternatively, you can take a shortcut and consider the quantum phenomena as they appear to you, without getting into the bad habits of physicists—their passion for complex theories and their joy in advanced mathematical analyses—and you can still get safely from point A (states of matter) to point B (states of numerical forms), understanding the essence of that process, even though you skip the technical details.

I have the feeling that the first option—getting a PhD in physics—is perhaps not such a good idea for most of us, at least for the time being. However, between the extremes of knowing a little bit and knowing everything, there is a middle road where you acquire sufficient knowledge to help you develop a useful understanding of things.

And so, back to Newton and his massy thing at point A. Newton's description of the appearance of particles can't be improved. They appear as solid, massy, hard, impenetrable, and moveable things. You can think of ordinary balls or round bullets; they like to push one another and, in collisions, they bounce off of one another. That is what Newton had in mind when he spoke of particles.

At the bottom of ordinary things we find the elementary particles or units of matter: the atoms and molecules and electrons. By definition, these elementary constituents of things differ in size from the things that they form, but it is generally expected that they are otherwise identical in essence. Thus, we expect that, like all material things, the elementary units of matter also are localized and compact; and we expect that they fill space solid and love to push and shove one another, and when they collide, they bounce off of one another. We see that, compliments of Parmenides, the ancient ideas can be quite helpful in forming an image of matter in our mind.

I don't know whether you have noticed or not, but the use of language in the last paragraph is a little bit tricky. I have said that *it is generally expected* that elementary particles are in essence like the larger material structures that they form, and we *expect* that they are localized and compact. I have not said that they actually *are* like ordinary round bullets and that the way in which they appear to us is all that there is to it. As it turns out, they are more than they show to us because, when they are on their own and out of sight, they cease to be material particles.

For the time being, point A has been taken care of. In everyday life all material things, even a beer truck, are in some sense lumps of matter. When you get hit by a speeding beer truck, your understanding of what the properties of matter are is instant! Now, to complete our task and to get to point B, we must consider the properties of waves.

Waves have properties that are in some sense contradictory or incompatible with the properties of particles. There are lots of everyday experiences of waves that show you what I mean. If you live on a coast nobody has to explain to you what waves are like. Likewise, rivers, lakes, and creeks all give us an instant understanding of what waves are about.

In all our experiences, waves are dynamic, playful, and creative, a moving sequence of constantly changing and shifting forms, an enjoyable dance, swinging up and swinging down in a fascinating game. The motion of waves is continuous; particles are discrete. Waves are extended in space; particles are localized. A wave is a dynamic process; a particle is a static point. If you play a musical instrument, you will appreciate the description of waves and particles by physicist and theologian J. C. Polkinghorne, when he says that waves come *legato,* while particles come *staccato.* From these comparisons one conclusion seems to be clear: One and the same thing can't have both the properties of waves and those of particles. *However, at the foundation of things, elementary particles, the electrons and atoms and molecules—the tiny bullets—do just that: Under certain conditions they act like particles; under other conditions they act like waves.*

When two waves meet in space, they don't bounce off of each other; rather, they embrace each other, dance, and prance around together. They merge their mountains and valleys; they interpenetrate and superimpose, constantly creating new shapes and forms. Physicists call this the *interference* of waves. Wave interference creates *interference patterns*.

Take a glass dish, as large as you can find in your kitchen, and fill it with water. Then, when the surface is calm, take a sharp object, such as a needle, dip it in and out of the water, and watch how the waves that you create spread out across the surface of the water in the dish, reflect at its walls, and run back and forth. Dip two needles into different ends of the dish and watch how their waves run into each other, superimpose, make large waves in some places and small ones in others, and form fleeting patterns. Waves are always on the run, spreading out in space, dispersing, and forming interference patterns. The beauty of this phenomenon has fascinated so many people that you can find many delightful interference pictures if you search the Web for the *interference of waves*.

Something special happens when a wave hits a wall that has a number of slits in it. In this case characteristic interferences are formed, because when waves pass through a system of slits, they break down into smaller wavelets that interfere with one another. (Even though the technical details don't concern us here, you may find it helpful to take a look at some of the pictures given in the appendix to this chapter on page 221.)

I think that, at this point, a short inventory of what we are dealing with is useful. We can use the following terms to describe particles: lumps of matter, pool balls, hard, fill space solid, discrete, noncontinuous, pushy critters. In contrast, the characteristics of waves: delocalized, extended in space, able to interfere and superimpose, coming legato. From this summary it is clear that, when a stream of bullets is shot through a system of slits, no fancy interference patterns appear: When bullets pass through slits, they form some simple piles behind them. However, here is the thing: *When elementary particles such as electrons, atoms, or molecules are shot through a system of slits, they form the interference patterns of waves.*

To sum up: At the bottom of material objects—beer trucks, bricks, and steel walls—we find elementary particles, tiny units of matter. They have all the properties of the larger things that they construct, but when we shoot them through some slits in a wall they create visible interference patterns, like only waves can do. This means that atoms and molecules, the components of all material structures, aren't only particles, but can also behave like waves.

Take a single atom. Say you could snip an argon atom out of the air and hold it in your hand. Argon atoms are real things; they have a definite mass and a constant size. In special microscopes we can see single atoms, and they always appear as tiny dots. However, when we take these dots and shoot them through a system of slits, the impacts of many of them will form an interference pattern. We have no other way to look at this than by saying that elementary particles can dissolve in states that have wavelike properties.

We started out with material particles and have now arrived at waves. Next stop: Waves to numbers.

PROBABILITY WAVES

Philosophers have a wonderful word—*entity*—that they use to denote something that exists, but which they don't want to specify exactly what it is. It comes from the Latin word for being, *ens*. It means just plain being, regardless of property. It is meaningful to use this term to describe what exists at the foundation of ordinary things, because the elementary units of reality can appear in states with incompatible properties that can't be described by a single name. We can't really say that we find elementary things—that is, material particles—at the foundation of material objects, because the entities we find there can also appear in the form of waves that don't contain any matter and have thoughtlike properties; these waves are more like *elementary thoughts* than *elementary things*. But we can't say, either, that we are digging up waves or elementary thoughts when we take things apart, because the entities that we find deep inside the visible world can also behave like ordinary material particles. So, to get out of this difficulty, we will simply use the abbreviation *ET* to denote the elementary units of existence at the foundation of things, leaving it open whether we mean *e*lementary *t*houghts (waves), *e*lementary *t*hings (particles), or unspecified *ent*ities. An ET is something that is there, but we don't exactly know what it is: ETs are what we find at the foundation of the visible world.

The ETs appear to us in different ways, which I will describe as different states. On the one hand, they can somehow acquire mass and appear as elementary particles. That defines one of the states that ETs can exist in: We can call it their *particle state*. On the other hand, ETs can also act like nonmaterial waves, and then their properties are so different from those of particles that they can't be in a particle state anymore: We will call this their *wave state*. This is like water. Water can be in a state in which it is a liquid and can make waves. Alternatively, it can be in a frozen state and appear as solid pieces of ice, but it is water all the time. It isn't the ETs that have contradictory properties; the states in which they can exist do. But there is no problem with that, because an ET doesn't exist in the contradictory states at the same time, like water isn't liquid and ice at the same time. Rather, an ET can jump from particle behavior to wave, and from wave behavior to particle.

The fact that microphysical entities can act in both ways, as waves and as particles, is often referred to as the *wave-particle duality*. In the older literature, the duality is often explained by saying that the form in which microphysical objects appear—waves or particles—depends on the mode of observation. If you choose an instrument that is sensitive to waves to observe them, you will record the properties of waves. If you choose an instrument that is sensitive to particles, you will record the properties of particles. In other words, the modalities of observing—or choices made by an observer—determine the modalities of being. To this I must say quite clearly: That interpretation is wrong. It is wrong because we never see the waves, only the effects of waves. It is wrong because microphysical objects *always* appear as localized particles when they are observed; they never appear as waves when we observe them. The presence of wave states is merely suggested by phenomena that are characteristic of waves. There is no duality; nature isn't schizophrenic. There is just one basic ET that can exist in different states.

The concept of duality is also unfortunate because it revives the Cartesian concept of the duality of the world as consisting of two disconnected types of things or substances: that is, mind (res cogitans or thinking stuff) and matter (res extensa or extended stuff). But the world isn't dualistic in the Cartesian sense. There is one reality, one type of ET, even though it can appear in different types of states. We don't speak of the duality of water because it can appear to us in the form of particles (ice) or waves (liquid).

From this it follows that the concept of *complementarity,* which is often used in this context, is also misleading. That concept implies that, for a complete description of reality, contradictory or complementary properties have to be used. But, like all objects of reality, microphysical objects have no contradictory properties. They simply can exist in different states that have different properties.

As described in the appendix accompanying this chapter, the so-called particle interference experiments are typically performed in a vacuum, where the interfering electrons, atoms, or molecules are largely isolated, and the interactions with their environment are at a minimum. Under such conditions, the transitions of ETs from their particle-state behavior to their wave-state behavior are spontaneous: *When electrons, atoms, and molecules are left alone, they become waves!* The wave states of ETs need privacy.

The details of how the ETs evolve in their wave-states are still a matter of research. For our purposes it is sufficient to realize that they can do it and do it spontaneously. As it turns out, all microphysical material objects behave like that: When they are isolated, the observable phenomena suggest that they dissolve spontaneously in extended fields of waves. In interactions with objects in their environment, these fields can contract abruptly to a point, which then appears to us as a material particle. So the Greeks, Parmenides, and Newton didn't get it right: Matter isn't enduring and lasting; it is a fleeting state, a shifty character, always ready to transform into waves.

An important question that comes up at this point regards the nature of the waves into which the ETs dissolve when they are left alone. They aren't water waves, or sound waves, or light waves. They aren't like any waves we know, so what are they?

We can shed some light on this question by first asking ourselves *where* a particle is when it becomes a wave. At first we are dealing with what we call a normal thing, an atom or a molecule, a localized mass that we see at a specific point in space. All of a sudden that thing flows apart as a wave that does what all waves do: It spreads out in space. So the question that comes to mind is this: Where *is* this particle when it is a spread-out wave? Has it become some sort of a skin? Could it be that the plastic wrap that you buy in the grocery store is just a single atom that some smart guy has caught on the wave and rolled into a box? Of course not: Particles always appear as localized events, not spread-out skins. So when an electron is in its wavelike state, where is it? What can we say about its position in space?

The answer that we must give to this question is simple: When an ET is in its wave state, when it is a wave that is spread out in space, *it is actually nowhere*. It has a definite position only in its particle state. In its wave state the space coordinates of the ET have no *actual* value but many *potential* values. This is so because the shape of the wave that an ET has become determines the probability to find it in space. This is a physical property of the waves into which the ETs dissolve. Physicists don't know *why* that is so, but they know that it is so and they take it as a given. It means, for example, that, at a crest of such a wave, the chances to find the ET in its particle state are high. Since these waves are spread out in space, crests can appear in many places. Thus, the chances to find a particle-turned-wave are very good in many places. This means that, when a material particle turns wave *it is actually nowhere*! It has no *actual* position in space, but many *potential* positions. It assumes an actual position only when it is a particle; that is, when it is in its particle state. The information contained in the wave states of ETs provides probabilities to find a particle in space. That probability is often referred to as *probability of presence*. The waves ETs dissolve into are *probability waves*.

We have watched an ice cube in a drink melt away into a surrounding medium. This is exactly how we can look at elementary particles when they become waves: They dissolve in a surrounding medium. The difference is that the medium in which ETs dissolve isn't a material medium in our space-time. There is no mysterious liquid there and no other stuff: the medium in which the ETs dissolve is empty of matter. When an ET enters a wave state, it abandons all matter. This follows from the fact that the waves into which the particles dissolve are probabilities. Because what are probabilities? Probabilities are numbers, ratios of numbers. Probability waves carry no mass or energy, just information on numerical relations. They *are* numbers, or patterns of numbers. *When they become waves, elementary particles become numerical patterns, mathematical forms, or numbers.*

And now we have finally reached point B: Matter has become numbers.

At this point, too, the thought-like aspects of ETs are coming to the fore: Probability waves don't carry any mass or energy, but are patterns of information. Information is something more thoughtlike than thing-like. Thus, you can't go very wrong when you think that thoughts play an important role in holding up your body and the things surrounding you. Information is usually intended to be used by a mind. The question is, whose mind?

COOPERATION AS THE CONDITION OF THE VISIBLE WORLD

How, then, do we explain the countless solid, lasting, and reliable things that surround us? This chair I sit on doesn't constantly flow apart in probability waves. The computer I write on is a stable thing, and so are the countless ordinary objects that surround us. Does their uninterrupted existence not show that the concept of ETs as mathematical forms is wrong?

The answer to this puzzle is simple: The tendency to evolve into nonempirical waves is a property of *isolated* elementary particles. When they aren't alone, they don't become waves and, except for very fleeting moments, they are material particles all the time. The atoms and molecules in ordinary things constantly interact with one another, and it is in these interactions that they find their empirical existence. The countless water molecules in a glass of water are constantly watching one another and keep each other straight. The atoms and electrons in your body are still constantly dancing in and out of the realm of forms, but only for such fleeting moments that they can't spread apart over large areas of space and tickle your nerves. For all practical purposes, they stay empirical all the time. An isolated particle is typically not a part of the visible world. It needs a partner or a community that supports it in its empirical existence. Visible reality is a cooperative phenomenon, or a relational mode of existence.

In an atom an electron is still a wave, but it has become a standing wave that surrounds the atomic nucleus. The interaction with the nucleus leads to a certain degree of localization. The localization isn't complete—the electron in an atom is still not a pointlike particle—but the probability to find it at large distances away from the nucleus is vanishingly small, albeit not zero. In the countless interactions that are going on in macroscopic objects, localization is practically, but not essentially, complete. And so the world of solid, lasting, reliable things goes on.

Empirical reality is a cooperative effect. It emerges in interrelations and isn't found in isolated things.

How the Concept of Potentiality Is Sneaking In

As we know from earlier, the waves into which particles dissolve are probability waves. As a wave, the material particle has no actual position in space but many *potential* positions. Thus, the wave states into which microphysical objects dissolve are *potentiality states*. When a material particle enters the realm of potentiality, it leaves the empirical world. You can't point at it and say, "Look, it is here," or, "Look, it is there." Such a thing *transcends* our experience; it is transempirical. It *transcends* the realm of matter; it becomes *transmaterial*. You can *think* numbers, but you can't see them.

Since these are the natural states of existence for microphysical objects—the states into which they will make spontaneous transitions whenever they can—we can conclude that the visible reality emanates out of a realm of potentiality that is underlying all things. Once emanated, it sustains itself as a visible world. It is in this way that we are led to the view that physical reality appears to us in two domains: the realm of the *actuality* of localized material things, and the realm of *potentiality* of the nonmaterial forms that are spread out in space. These forms are real, even though they are invisible, because *they have the potential* to manifest themselves into the empirical world and act in it.

The actuality—the empirical particle—emerges out of its state of potentiality when it interacts with objects in its environment. This rule seems to apply to everything: The visible world is an actualization—an emanation—out of a domain of transmaterial and transempirical potentiality forms. *The basis of reality is a domain of transmaterial forms, images, or elementary thoughts.* Physical reality is driven by inner images, like a human being is driven by the images in the mind.

Regarding the mechanism by which ETs make the transition from wave states to particle states, it is now thought that the properties of matter emerge in the interactions of ETs with an invisible field, called the *Higgs field,* that permeates the universe. The possible discovery of a specific elementary particle called the *Higgs boson* in recent experiments performed at CERN, the European Organization for Nuclear Research, has given this theory new support.

If you think that this view of things is hard to accept, many pioneers of quantum physics will applaud your reaction. Niels Bohr, for example, never accepted that the wave functions of quantum theory are in any way real. The wave functions can organize our observations of things and they allow us to make predictions, he said, but that is it. In his view, the quantum waves are simply some tool or a useful formalism. If you want to put it into some fancy terms, you can say that the wave functions of quantum physics aren't ontological, but epistemological entities: They are elements of theory, but not of the physical reality.

In some of his writings, Werner Heisenberg offered a similar view: The wave functions of quantum theory, he said, represent our knowledge of things. David Bohm, another pioneer of quantum physics, thought that particles and waves are real and coexist at all times, and the waves have the function of pilots that guide the actions of particles. It was only relatively recently that British physicist C. N. Villars proposed that the wavelets whose actions we encounter at the quantum level are truly existing entities, which he called *potentiality waves,* because they contain the potential of a system for future empirical events. In Villars's description, potentiality waves "are conceived as physically real waves which exist in their own right, not merely as representations of the behavior of particles. Microphysical objects are not particles 'guided' in some mysterious way by 'waves of probability,' but, rather, microphysical objects *are* waves of potential observation interactions." A potentiality state is generally defined by the fact that the physical properties of a system in such a state, as Villars described it, don't have an *actual* value, but many *potential* values. That sort of situation isn't restricted to the position in space but can also apply, for example, to the energy of a particle or its direction of motion. As far as position is concerned, when a thing has no specific place in space, it can't be a part of the empirical world. This aspect holds for all potentiality states: When an ET enters the realm of potentiality, *it leaves the empirical world;* it enters a state in which it transcends human experience; it becomes transempirical; it doesn't exist, as Villars describes it, "in ordinary three-dimensional space." The empirical particle emerges out of its potentiality state when that state interacts with an object in its environment. In that process, the potentiality state necessarily ceases to exist. This establishes the nonempirical character of potentiality states: *Observations destroy them.* "During the act of observation," Werner Heisenberg wrote, "the transition from the possible to the factual takes place." Thus, potentiality states are necessarily nonempirical because, when we take a look, they disappear. That they must be invisible also follows necessarily from the fact that they are empty—they carry no mass or energy. We can't see empty numbers.

Potentiality waves are invisible not only because they are numbers but also because they represent a state of waiting to be actualized. In the actualization they cease to exist.

Visible objects are always somewhere. Even though you don't know me, you know that, while I am alive, I must be at some point on this globe. At that particular point the probability to find me is

1, or 100 percent. I can't be in a state where the probability to find me is 10 percent in my home, 10 percent in the pub down the street, 10 percent driving in my car, and so forth. This isn't the case for ETs that exist in a network of potentialities.

It seems that the interplay between actuality and potentiality in physical reality is quite general: *The visible world is an actualization—an emanation—out of a realm of transempirical and transmaterial potentiality.* We don't know with certainty the nature of the forms in the realm of potentiality—after all, they are invisible—but physics posits that they are waves. It might be more precise to say that the forms of the realm of potentiality interact with the world as though they were waves. There are indications that the waves in the realm of potentiality are hanging together, like the water waves in an ocean, so that reality is an indivisible wholeness—the One—in which all things are interconnected.

With this we have arrived at an important goal of our journey. The basis of the visible world is an invisible, nonmaterial wholeness. Reality appears to us in two domains: the visible world of material things and the invisible realm of potentiality. The visible world is an emanation out of a realm of forms. When material particles dissolve in potentiality waves, they cease being material particles. Instead, they transform into something that we can describe only as invisible mathematical forms—numerical shapes—that transcend the realm of matter. The basis of the visible world is a realm of hidden images.

The emergence of wavelike properties in the behavior of elementary particles forces us to accept some amazing conclusions regarding the nature of physical reality. There is a realm of the universe that has the nature of potentiality—a realm that isn't made up of visible, material, and energetic things, but of invisible mathematical forms: patterns of information or images. People who aren't comfortable with a transempirical and nonmaterial reality often argue that many interpretations of quantum theory exist that are all possible and yet contradictory. The idea of a transempirical reality, they say, is yet another meaningless interpretation. It is true that contradictory interpretations of quantum theory exist, and experiments can't distinguish between them. However, these contradictory interpretations all share one common feature: To explain the visible world they all employ aspects of reality that are nonempirical.

Potentiality is a physical state of the universe: invisible, transempirical, and transmaterial but *powerful* and *real*. In an empirical science that claims that the world can be understood by observing and measuring its visible surface, the discovery of the realm of potentiality was a shock. It marked the end of the era of Newton's and Darwin's materialism and the dawn of a new idealist era that seeks the essence of the world in a transcendent part of reality. For the understanding of your own nature the discovery was a triumph: Since a realm of potentiality exists in the universe, it can also exist in you as an inner potential that takes you beyond the limited possibilities of your material body and brain.

THE EQUIVALENCE OF MATTER, ENERGY, AND POTENTIALITY

It is a logical challenge to the view that material particles can enter a state of potentiality waves to ask this: Where does the mass go when an elementary particle turns into a potentiality wave? How can something tangible and visible turn into something invisible?

As it turns out, the principle isn't really new. Einstein already showed that matter can turn into something non-material and invisible—that is, into energy. But, what is energy?

When I ask my students this question, they typically answer, "Energy is the ability to do work," because that is the definition of energy offered by physics. The problem is that this defini-

tion doesn't tell us what energy *is*. It merely replaces one term—
energy—by another—*work*. But I don't know what *work* means.
So my students reply, "Work is something that you do when you
change the state of motion of a mass." Like, energy has to be spent
to accelerate your car. When your car starts to move, it has to do
work! Very good! But does that tell us what *energy* is? No! All that
this answer does is to replace, once again, one term, namely *work*,
by another, namely *mass*. So, I have to ask, what is mass? At this
point Einstein's definition comes to mind: Mass, he said, is energy.
Thus, the answer to our question "What is energy?" is the wonderful insight that energy is energy.

Einstein's famous formula tells us that when mass is multiplied
by the square of the speed of light, you get energy ($E = mc^2$). The
formula isn't usually interpreted in this way, but it tells us that matter, something solid and tangible, can spontaneously turn into energy, something non-solid and intangible. So, what is the big deal
in thinking that mass can turn into potentiality? It is the same sort
of process. The conclusion is that there is no logical difficulty with
the intangible states of potentiality: Matter can turn into the forms
of potentiality in the same way in which it can turn into energy:
This is the equivalence of mass, energy, and potentiality.

DOES CONSCIOUSNESS CREATE REALITY?

In connection with the collapse of the wave state of an ET to its
pointlike particle state, it is often said that "observation creates reality" or "reality is created by our consciousness." The idea goes
back to suggestions first made by the great Hungarian pioneer of
quantum theory, John von Neumann. I think that such statements
are wrong or misleading for two reasons.

First of all, no consciousness is needed to collapse the wave state
of an ET to its particle state. The collapse is a consequence of a

physical interaction of an ET with other ETs in its environment. It is interaction that puts an ET out of a state of potentiality into a state of actuality. We can then have information on what that state is, but that doesn't mean that the information has *created* this state. The fact that information is possible is a consequence of the ET occupying a particle state. Furthermore, no involvement of consciousness is needed to achieve the transition from the state of potentiality to the state of actuality. The ETs can do it on their own.

A second misunderstanding is that, in the process of actualizing an ET into a particle state, *reality is created*. This is completely wrong. It is the mistake of equating "being real" with "being visible," or "being real" with "being material." ETs are always real. The invisible, nonempirical, and nonmaterial forms in the realm of potentiality are a part of reality. When an ET is in its wavelike state of potentiality, it exists. It is always there; it is simply not visible. If it didn't exist, it wouldn't have the potential to appear in the empirical world.

Interaction can actualize a material structure in the visible world, but it doesn't create it. The potentiality is already there and real before something actualizes out of it. The contrast between potentiality and actuality isn't between the real and the nonreal; it is between the visible and not visible, or between empirical and transempirical.

The notion that observation creates reality is a striking example of how deeply rooted convictions, such as the classical empiricism, can brainwash even the most brilliant scholars.

POTENTIALITY AS A PHYSICAL STATE OF THE UNIVERSE AND OF YOU

Physicists find it completely plausible to suggest that the universe is filled with matter, perhaps with dark matter, or that there is a cosmic field of electromagnetic waves, like an ocean of energy. If that is so, we must think that the universe is also an ocean of potentiality, because wherever matter and energy appear, the aspects of potentiality also appear.

And it is at this point that we realize that, if potentiality is a cosmic property, we have to accept that it is also active inside us.

If something invisible is of such importance for our life, we have to determine whether it really exists, or whether the concept of potentiality is just one more operational principle that helps us to handle the quantum phenomena, but is otherwise meaningless. Thus, the question is this: Is there any visible evidence that an invisible reality exists in this world? The question seems to answer itself: How can there be visible signs of an invisible reality? However, empirical evidence does exist that shows that the properties of atoms and molecules depend on invisible empty states, which quantum chemists call *virtual states*. Thus, the next step in our journey is to take a look at what virtual states are and why we have to think that they are real.

Chapter 2

YOUR POTENTIAL IS REAL EVEN THOUGH YOU CAN'T SEE IT: HOW VIRTUAL STATES ACT ON THE VISIBLE WORLD

"Not all potentiality is converted to actuality in any finite time. There are innumerable clouds of probability running around in the universe that have yet to trigger some registered event in the macroscopic world. We have every right to assume that the universe is filled with more uncertainty than certainty."

—JOHN ARCHIBALD WHEELER WITH KENNETH FORD

The other night I had a strange dream. I dreamed that I was invited by Dr. Erwin Schrödinger to come to his house. The way things go in dreams, his house in Austria was just a few blocks down the street from my house in Fayetteville.

I arrived at the requested time, and his assistant, a beautiful young lady, opened the door. "Herr Doktor is waiting for you in the living room," she said and smiled at me. "Down the hallway; second door to the right."

When I entered the room, I froze in shock. Dr. Schrödinger was changing a lightbulb in one of his ceiling lights and he hovered high above the floor, but I couldn't see a ladder on which he was standing. He was standing on something that looked like the step of a ladder, but there was no ladder; the step was floating in the air. "Oh, my God!" I thought, instantly in panic. "That man is going to fall."

"Dr. Schrödinger!" I cried out. "What are you doing? For heaven's sake! Be careful! You are going to fall."

Smiling, Dr. Schrödinger looked down at me, calmly screwed his bulb tight, and said, "Ah! There you are! Welcome to my home, and make yourself comfortable. Everything is fine! I will be with you in a second."

"But!" I said, scrambling for words. "But . . . but you are floating in the air! You are going to hurt yourself. Let me call your assistant and we will get a ladder to you, immediately."

"But I am standing on a ladder," he said. "Don't worry! You may not see it, because it is a virtual ladder, but they are the best."

"A virtual ladder?" I said. "What do you mean, 'a virtual ladder'?"

"Virtual ladders are ladders whose steps you don't see, except for the one you are standing on," he said. "But all the steps are there. Trust me—they are real! As soon as you touch one of them, it becomes visible. You see, I am standing on one!" And with this he started to come down from the ceiling, step by step. Sure enough, every time his foot reached down, all of a sudden a step appeared, and at that same moment the step on which he had been standing vanished.

"Wow!" I said when we shook hands. "That is impressive! How did you do that?"

"I didn't have to do a thing," he said. "This is how nature works. It is full of virtual states. These states are like the steps of that invisible ladder—as soon as you touch one, it appears in the visible world."

He led me to a chair and reached for a bottle of wine. "I have a good Austrian Riesling here," he said. "May I pour you a glass?" He took out the cork and made preparations that looked like he was going to pour the wine on the plain table.

"Excuse me, Dr. Schrödinger!" I said. "I would prefer to have mine in a glass. There is no glass! Don't waste your good wine by pouring it on the table!"

"There is a lot that you have to learn," he said. "Of course there is a glass! But, since it is empty, it is a virtual glass. It is in a state of potentiality and you can't see it, but it is there. All you have to do is to fill it, and, bingo—to your health!"

And with this he poured the wine onto the table, but just before it hit the surface, a glass appeared. "The world is full of potentiality," he said. "Potentiality is invisible, but you can make it visible! It is part of its nature to be invisible. That gives it potential! But all that you have to do is to reach out for it, and it will become real for you. That is my message to you today. I may contact you again in a few days! Have a good day!"

His last words started to sound a little bit blurred. When I looked up, he seemed to flow apart like a wave and I decided that the time had come to wake up.

Meet the Virtual States of Atoms and Molecules

As dreams go, this one was a mixture of fact and fiction. The fiction, of course, is that a person can stand on imaginary ladders with invisible steps, and that an ordinary object, such as a glass, will vanish from the visible world when it is empty. However, the fact is that all atoms and molecules contain energy ladders with countless real but invisible steps. We could even say that atoms and molecules *are* energy ladders in which some of the steps are visible, while others aren't. Physicists call the steps of these energy ladders the *states* in which the atoms and molecules exist, and quantum chemists call the empty states *virtual states*. The virtual states are important, because they determine what atoms and molecules can do in this world in the future. We call the future possibilities of a system to act in the visible world its *potentiality*. In this chapter we'll take a look at the virtual states of atoms and molecules, because they are models for the potential that is hidden in each one of us and determines what we can do in the future. The best way to enter the world of virtual states is by taking a look at the simplest atom of the world, the hydrogen atom.

A hydrogen atom consists of two elementary particles, a proton and an electron. The proton has a positive electric charge and nearly all of the mass of the atom, and it forms what is called the *nucleus* of the atom. The proton is about eighteen hundred times heavier than an electron. The electron has all the negative electric charge of the hydrogen atom, and it resides somewhere in the space surrounding the nucleus. The question is, where and how?

In 1913, Danish physicist Niels Bohr developed the *planetary model* of the atom. He postulated that electrons in atoms move about the nucleus in orbits with fixed radii, like the planets revolve around the sun. Each radius corresponds to a fixed amount of energy that defines an energy level, which is called a *quantum state* of the hydrogen atom, because physicists call any fixed amount of energy a *quantum* of energy. Thus, the quantum states of the hydrogen atom form a natural energy ladder, and when the electron is in a specific orbit around the proton, or so Bohr thought, it occupies one of the steps of the ladder. It must always occupy one of the steps; it can't hover between two steps. That is like when you use a ladder to climb up on the roof of your house: You have to stand on one of the steps of your ladder; you can't stand in between two steps. At any given moment one of the energy states in the hydrogen atom is occupied, while the others are empty. This is why physicists speak of *occupied states* and *empty states*. In this model of the hydrogen atom, the radius of an occupied orbit can presumably be measured by watching the electron as it moves around the proton, but the empty orbits are, of course, invisible, because there is nothing there to see.

Bohr also proposed that transitions between the steps of the ladder are possible, in that the electron can jump from an occupied state to an empty state. In these jumps, energy is exchanged between the atom and its environment in the form of the energy quanta of light. Physicists call a quantum of light energy a *photon* (see the appendix to this chapter on page 245). When an atom jumps from a lower energy state to a higher one, it has to absorb a photon from its environment with exactly the right energy to pay for the transition. When it jumps from a higher state to a lower one, the energy difference is set free and a photon is emitted to the environment of the atom. Since the jumps are between quantum states, they are called *quantum jumps*.

It was a triumph of Bohr's theory that it was able to explain completely the properties of hydrogen atoms. However, it couldn't explain anything else: It couldn't explain the heavier atoms and it couldn't explain the properties of molecules. Some other theory had to be found, especially once it was discovered that the electrons in atoms keep the nuclei at a distance that is tens of thousands times greater than the size of a nucleus.

When this aspect of the structure of atoms was discovered, it was an enormous challenge to the physicists, because it violated some fundamental rules of classical physics. Just think about the opposite electric charges of electrons and protons, and how oppositely charged particles attract each other, like two magnets attract each other. When two particles with opposite electric charges get close to each other, they instantly crash into each other. Thus, according to classical physics, an electron in a hydrogen atom shouldn't float around the proton, and the two shouldn't stay apart but instantly snap together. This applies to not only the electron in the hydrogen atom; all electrons in all atoms should crash into the nuclei.

However, floating around the nuclei and staying away from them is what the electrons in atoms do. We know that they are doing this because they create an empty space that is huge by atomic standards. In the Introduction, when we explored what is really real, we already talked about the immense empty spaces in all things. For example, if the nuclei and the electrons of all the atoms in your body could be stacked together, you would shrink to a microscopic point. So, the discovery that the electrons in atoms stay away from the nuclei was an incredible challenge to the physicists in the early 1900s. The only possible solution to this enigma was that the electrons in atoms aren't ordinary material particles, but something else. The question was, what could that something else be?

In the previous chapter we saw how material particles turn into waves when they are left alone. In atoms, the ETs aren't alone, but could it be possible that the electrons in atoms are waves, nevertheless? And could it be that, by becoming nonmaterial waves, like ETs do, the electrons in atoms aren't subject to electrostatic forces like ordinary particles are?

Something like this must have gone through Erwin Schrödinger's mind when he developed his version of quantum theory, which is called *wave mechanics*. To solve the problem of the atom, in 1925 Schrödinger proposed that, when *an electron becomes a part of an atom, it is no longer an enduring material particle but a wave.* Electrons in atoms can exist in states in which they are standing waves. Being a wave allows an electron to stay away from the nucleus in a natural way. This is so because Schrödinger's waves aren't matter or energy waves but probability waves; that is, they are numbers or ratios of numbers, or numerical patterns: *In the wave states of atoms the electrons are probability waves.* By this we mean that the squared amplitudes of these waves tell us where an electron can be found in the space surrounding the nucleus. The probability to find the electron-turned-wave in an atom or molecule is often called the *probability of presence,* or the *probability density.* Since they describe what used to be an orbit in Bohr's theory of the atom, the squared wave functions are also often referred to as *orbitals,* meaning something like an orbit.

When I say that electrons in atoms are waves, I'm using a short term for a very complex situation. Precisely what I mean is that an electron in an atom can dissolve in wavelike states, which are like the waves of probability or potentiality that we encountered in the double-slit experiments (see chapter 1 and its appendix). In these wave states, an electron is still there, but not as an empirical material particle: It has become a form of probability or potentiality to be found. Like before, the electron-turned-wave can pop out of the state of potentiality and appear in the empirical world as a particle with its mass and charge, but then it will instantly dissolve again in its wave state.

If you could crawl into an atom and look at the landscape surrounding the nucleus, what would you see? You would see nothing! The landscape surrounding the nucleus is filled with potentiality, and you can't see that potentiality. Once in a while—bingo—you can see an electron popping out of the potentiality as a little dot, and the shape of an orbital appears in repeated observations of this kind. This is like the shape of an electron interference pattern in a double-slit experiment that appears in countless single-electron impacts. We should add that, even though they are invisible, the shapes of orbitals are immensely important. They determine, for example, the structures of molecules. If the electrons in atoms were nothing but solid material dots, the molecules that we know wouldn't have the structures that they have, and you wouldn't be around to know it, because the structures of the molecules in your body determine your biochemistry. All of these aspects and activities of electrons have to be kept in mind when we say "electrons in atoms are waves." By the concept of a "standing" wave we mean waves that don't flow apart and propagate and spread out in unlimited space, like the waves that ETs form in a vacuum. The standing waves that electrons form in atoms are somewhat concentrated or localized in that they swing up and down around the nuclei in stable patterns. At the level of ETs, interactions localize.

As a probability wave, an electron isn't a part of the empirical world—probabilities are numbers, and you can't see numbers—

but it has the potential to pop out in it. This is why I said that the probability waves into which the electrons turn are, at the same time, also potentiality waves. This means that the largely empty space that surrounds the atomic nuclei isn't really empty; it is empty of matter, but *filled with potentiality.*

When two atoms collide in space, their potentialities collide and may form a molecule. When a molecule is taken apart, the identically same atoms will reappear that formed it, with their nuclei, electrons, and potentialities all unharmed and completely restored. How strange this is! If two solar systems with their planets and suns crash into each other and form a single planetary system, will the same solar systems that existed before the crash come out of the mess, when the two are somehow pulled apart? Not likely! It is like Heisenberg said: "Atoms do not exist as simple material objects."

So, in a nutshell: When an electron and a proton form a hydrogen atom, the interaction leads to the formation of countless atomic states, or quantum states, that the electron can occupy. Each state is defined by its fixed amount of energy and the specific form of the wave that an electron assumes when it occupies this state. Since the electron in a stable hydrogen atom occupies only a single state at any given time, the atom is a system of countless empty, or virtual, states. Virtual states are real but invisible, because there is nothing there to see. But they are there all the time, nevertheless. If an empty state weren't there all the time, an electron couldn't jump into it. We could also say that, if the empty states didn't exist as a possibility in an atom, they couldn't be actualized. And the specific probability patterns of the empty states must also exist all the time, or they would have to be created out of nothing when an atom took a jump into such a state. The probability pattern of an empty state isn't an observable pattern, because there is nothing there for which the probability of presence could be determined. Thus, *the virtual states of a hydrogen atom are invisible, but real.* This rule applies to all atoms and to the molecules that they form.

We said above that, in addition to being states of probability, virtual states have the nature of potentiality. This is so because the virtual states contain the future empirical possibilities of a system. A molecule, for example, can't do anything but actualize one of its virtual states. It can't become anything but what is already contained in it as a possibility. It can jump from an occupied state into an already existing virtual state; it can't jump into nothing or actualize nothing. We hear echoes here of Pythagoras and Plato, the Greek philosophers, who thought that everything comes out of a realm of forms. The virtual states are empty of matter but *filled with potentiality*. They represent the possibilities of the material world.

This structure of the world is so important to understand because your own situation is very similar. You, too, have a future only because of the inner potential in you. You have an inner potential because everything in this universe has an inner potential. It wouldn't make sense to think that the structure of your mind is the only exception and alien to the rest of the world. Much of the unhappiness in this world is due to the fact that many people aren't aware of their inner potential and so do nothing about it, because it is hidden deep inside in invisible states and easy to neglect. The virtual structure of things is important at the human level, because it is a model of our personal structure.

DREAMS OF VIRTUALITY

I couldn't get Dr. Schrödinger's assistant out of my mind. I couldn't stop wondering whether she was real or virtual and waiting to be actualized. Perhaps, I thought, all of us need to be touched by another human being to become real in the human world.

In any case, it didn't take long before I had another invitation and another dream: Dr. Schrödinger asked me to see him again.

So, flowers in hand, I walked down the street, crossed the border into Austria, and went straight to Dr. Schrödinger's house. Before I could ring the bell, the door jerked open and a grouchy old butler appeared and sneered at my flowers. "Down the hall!" he said grumpily. "Second door on the right!"

"I am glad you could make it!" Dr. Schrödinger greeted me. "I have something important to show and to give to you.

"I know that students usually have a hard time understanding what the virtual states in atoms are. I am glad to help and want to give you a special device, a chest of drawers—actually a chest of quantum drawers—that you can use to demonstrate the nature of quantum states.

"It is very important for students to understand," he continued, "that virtual states are real. It is true that virtual states are mere mathematical forms and invisible patterns of information. I am quite proud of this discovery. Nevertheless, these states are real, and that is important to understand, because the potential in you is of the same kind: It is invisible and yet it is vital, and if you aren't aware of your inner potential, it won't ever get actualized. And what a waste that is."

I nodded, quietly agreeing with him. I know a lot of people who lead frustrating lives, because they don't live up to their true potential, thinking that actualizing the potential of eating and sleeping is enough.

"The empty states of quantum systems aren't mere formulas of mathematical shapes," Dr. Schrödinger continued, "because they have the potential to act in the visible world. Thus, they are some truly existing entities. A material system could do nothing—I repeat: *nothing*—if it didn't have virtual states. As a matter of fact, without the potential in you, you could do nothing that would be worthwhile to live for! If you deny that the virtual states of atoms are real, you must deny your own potential at the same time, and you may never find the fulfilling life that is possible for you!"

With that Dr. Schrödinger walked to a piece of furniture that looked like a nightstand. It had a single drawer that was standing open, and there was a rabbit in it, hopping around like crazy. "Take a look at this chest of drawers," Dr. Schrödinger said. "I have prepared it especially for you."

"Chest of drawers?" I said. "I don't see any drawers. Just a single drawer and this thing jumping up and down in it."

"First of all, this isn't a thing," Dr. Schrödinger said. "It is an entity of life—you can call it an EL—and a good way to look at ELs is to think of them as elementary quanta of life. This particular EL appears as a rabbit in this particular drawer. In other drawers of this chest, it will appear in a different form.

"Furthermore," he continued, "when you say that you 'don't see any drawers,' that is exactly my point. This is a special chest of drawers: a chest of quantum drawers made to be occupied by ELs. You don't see these drawers when they are empty, but they truly exist, nevertheless! Right now, only one of the drawers is occupied. It is the one where an EL is given the form of a rabbit. But invisible things can be real, and this is a chest of real drawers with all their properties. They are open and ready to get occupied by an EL, or quantum of life. Just watch me! As soon as I clap my hands, this EL will jump out of the rabbit drawer and into another one. When that happens, the drawer in which it is now will become invisible, and the drawer in which it will jump will become visible!

"But before I do it," he said, "I want you to take another look at this particular EL."

All along, while we were talking, the EL had been hopping like crazy in its drawer, and while I was thinking that having a thing like this around me all day would make me nuts, Dr. Schrödinger suddenly clapped his hands and the beast took a screaming leap five or six feet up in the air! While it was airborne, it blurred out of sight and the drawer that it had left vanished. All of a sudden, four or five feet away, another drawer appeared with a crashing sound, and the rabbit had turned into a chicken.

"Wow! That is wonderful!" I said. "I don't really like to eat rabbits, you know, but I do like fried chicken. And, you are right, my students would certainly love to watch this experiment."

"Well!" Dr. Schrödinger said with some pride. "This set of drawers isn't exactly intended as a kitchen utensil. It is a highly complex scientific instrument that I designed to demonstrate what we mean by the 'probability of presence' of a quantum object.

"For example," Dr. Schrödinger continued, "the probability of presence of a quantum of life that is put into this particular drawer is that of a chicken. But now, watch this!" He clapped his hands again, and instantly the EL took another screaming leap and vanished from sight as its drawer disappeared. In another part of the room a new drawer appeared and a squirrel danced right and left around in it. "Now, what do you think?" Dr. Schrödinger asked.

"Dr. Schrödinger," I said, "I am impressed! But may I ask you a question? Last time when I came here I was received by the most charming and friendly young lady! When she opened the door, she had the most wonderful smile for me. But today? That grouchy old man? Is that what happened to the young lady? She made a transition into another drawer?"

"Don't get involved with such details!" Dr. Schrödinger said. "I can't control that girl! If she wants to occupy a state in which her personal ELs appear as an old man, there is nothing I can do about it."

"You mean," I responded with some astonishment, "it is that easy? All that we have to do to change our personality is to find the right virtual state for the ELs in us and then make them take a jump in it?"

"I don't want to discuss these things right now," Dr. Schrödinger said. "But I want you to take this drawer and show it to your students. I have constructed it in such a way that, in addition to rabbits, chicken, and squirrels, there are states in which the probability of presence of the ELs is that of a kitten, a dove, a mouse, and many others. Be careful when you happen to hit on the drawer of the rat. That potentiality can be a problem. I hope that your students will learn from this instrument that the virtual states also exist in their minds and that the most wonderful things can come out of them and make this a better world."

With this he looked at his watch and said, "Get your drawer right now! It is actually easy to carry, because all of its drawers, except for one, are empty. But we have to hurry, because my time is running out."

Like with everything else, Dr. Schrödinger was right in that regard, too. I was excited to accept his present, looking forward to sharing it with my students. But while I was thanking him, all of a sudden the entire house—Dr. Schrödinger, the quantum drawer, and everything else—was flowing apart into some fuzzy mess. Before I knew it, everything was gone.

"Rats," I thought. "That was a true opportunity! Or, should I say, a true potentiality?" The thought went through my mind that some potentialities in us are like stationary quantum states; that is, they exist for a long time. Others are like the fleeting forms of transition states, which appear and are gone before you know it.

When I got home, some squirrels were chasing each other, hopping up and down our oak tree. "You can't fool me!" I said to them, and was happy to wake up.

How the Virtual States Show Themselves in Chemical Reactions

Like before: fact and fiction. The quantum properties of atoms and molecules don't apply to macroscopic objects. There is no home furniture whose drawers become invisible when they are empty. And who would want it! But the atomic and molecular quantum states are like drawers that vanish from the empirical world when they become empty, and yet they continue to exist in a system as a part of the real world.

Similarly, ordinary objects don't change their forms when you take them out of one room in your house and put them into another. But atoms and molecules do just that: When they change their quantum states, the waveforms that determine their physical and chemical properties change. It isn't only that electrons in different states are associated with different forms. No, they *are* these forms. The equivalent at the level of ordinary things would be the change of a rabbit into a squirrel when the rabbit hops off the lawn onto an oak tree, and the change of the squirrel into a hummingbird when it jumps from the oak tree into a blooming bush, and so on.

When we claim that invisible states exist that are real, we have to come up with some visible evidence, some phenomena, that support that claim. As it turns out, many such phenomena can be found, among them the chemical properties of atoms and molecules.

When molecules interact with one another, they can start a chemical reaction. In a chemical reaction the reactants are transformed into products; that is, they become new kinds of molecules. The way chemical reactions proceed depends on the chemical properties of the reacting molecules, which, in turn, depend on the virtual states of the reactants. If the virtual states of a molecule are all high-energy states, they are difficult to reach, and such a molecule is chemically inactive. In contrast, when the virtual states of a molecule are lying low on its energy ladders, that molecule is highly reactive. Thus, virtual states affect an empirical property—that is, the chemical reactivity of molecules.

There is a special class of reactions—chemists call them *redox reactions*—in which reacting atoms or molecules exchange electrons with one another. This exchange can lead to measurable changes in magnetic properties, which depend on the virtual states of the reactants.

A particularly dramatic example is found in the reactivity of oxygen. Breathing oxygen can serve your metabolism, because oxygen molecules contain a special category of virtual states, which chemists call *degenerate states*. Because of these degenerate states, oxygen molecules are particularly reactive and can oxidize your food. In contrast, the virtual state structure of the so-called noble gases is such that they hardly react with anything. This is why they were called *noble* to begin with: Nobility doesn't mingle with ordinary folk.

Many other examples can be found that demonstrate that virtual states are real, even though we can't see them, because they affect observable chemical reactions.

How the Virtual States Show Themselves in the Interactions of Molecules with Light

Skyscrapers are real, and their elevators can take you up and down in a building. When you take an elevator, it allows you to go from one floor to another, but you can't get out in between. You can press the button to go from the first floor to the second, or to the tenth, but there is no button to stop at floor number two and a half. That sort of thing is just not done!

That is exactly like the quantum jumps that take atoms and molecules from one step of their energy ladders to exactly another and not in between. The only difference is that the floor levels to which an atom can jump aren't visible before the atom gets there. The equivalent situation in a skyscraper would be scary, in that the floors would be visible only when some people were on them. An empty building would be invisible entirely, and you would have a hard time finding it. In a building in which some people are on the thirtieth floor and some are on the tenth, while all the other floors are empty, you would see the two occupied floors floating in the air and expect them to come crashing down any moment. But empty quantum states never crash.

If we keep track of the number of people who visit a certain building on any given day and where they are going, we can determine the relative frequency with which the different floors are visited. These frequencies may vary considerably. For example, a popular restaurant on the fifth floor may attract more visitors every day than a shoe store on the seventh floor. In that case, the transition probabilities between the ground floor and the fifth and seventh floors would differ significantly.

For molecules we can do exactly the same thing. Molecules usually reside in their ground states. The ground state is the state with the lowest energy. From there they can "visit" higher lying states, provided they have the energy to do that. They can get this energy, for example, out of a light beam that is shining on them. Using special types of instruments, called *spectrometers,* we can determine how many out of a given number of identical molecules will make transitions between their ground states and various higher lying states. Such numbers define the relative *transition probabilities* between the quantum states of a molecule.

As in the case of a building with different businesses on different floors, the transition probabilities between the different states of a molecule can vary significantly. Differences in transition probabilities mean that the possible state transitions in a system aren't equally probable, even if enough energy is there to provide for all of them. Independent of the availability of energy, in all molecules there are some state transitions that seem to be easier than others. And some are never observed: Spectroscopists say that such transitions are forbidden. As a consequence, each molecular state transition is characterized by the probability with which it can occur.

Transition probabilities can be calculated precisely, if the wave functions of the quantum states of a system are known. As it turns out, such probabilities depend on the wave functions of *all* the states involved in a transition, including the empty states. *Thus, the waveforms of invisible empty quantum states determine observable quantities—that is, the observed spectroscopic transition probabilities of atoms and molecules.*

Prior to making a jump into an empty state, an atom or molecule feels out, so to speak, how suited a given state is for a quantum jump. To do so, the system has to leave the empirical world for a short time, enter the virtual state space, and inspect the empty state that it can jump into. Whether a jump will actually occur depends on the waveforms of the virtual states. *Thus, the invisible logical order of virtual states can affect the outcome of visible phenomena: That is, virtual states are real.* (For more on electronic states and transition probabilities, see the appendix for this chapter on page 245.)

THE MINDLIKE ASPECTS OF VIRTUAL STATES

State transitions in atoms and molecules are comparable to the events that your mind goes through when it receives a signal from your senses. When you have to respond to an external stimulus, your mind will first reach into the unconscious realm of inner images and then it decides which one will be the best to actualize.

The virtual states of atoms aren't only nonempirical forms; they are also forms of potentiality—that is, they have the potential to express themselves in the empirical world when they get occupied. From this we see that the concept of potentiality is indispensable when we want to describe what microphysical systems are like.

This brings us to the thoughtlike aspects of virtual quantum states, because our thoughts are also nonmaterial, and nonempirical forms that have the nature of potentiality. The thoughts in you exist, but not as empirical structures. Nobody can look into your brain and watch a single thought. At the same time, your thoughts are real for you, because you know that they are in your mind and you can think them. They exist in you as a potentiality because they *can* express themselves in the empirical world; for example, you can express a thought in spoken words or write it down on a piece of paper.

Thoughts belong, of course, to a mind. It is in this sense that I call them mindlike. They aren't matterlike or energylike, but mindlike. Does that allow us to conclude that, at the level of atoms and molecules, mindlike order expresses itself spontaneously in the material world? Certainly, apart from what we can factually say at this point, nothing can stop us from exploring this possibility.

All the states of atoms and molecules are patterns of numbers, information, or wave functions. But, as we have seen, we have to distinguish between empty states and occupied states. They differ in that occupied states have observable properties, while empty states are invisible patterns of logic. Even occupied states aren't visible in the manner in which the solid forms of bricks and steel balls are visible at a single glance, but their patterns appear in repeated measurements. You can think that the patterns of occupied orbitals can be made visible like the single-particle interference patterns that are discussed in the appendix of chapter 1. That is, each observation of an occupied orbital will produce the dotlike appearance of an electron, and the pattern of the orbital will appear when the dots of many observations merge together.

The relationship between transempirical virtual orbitals and empirical occupied orbitals allows us a glimpse at how something thoughtlike can express itself in the empirical world. The first step of this process lies in the transformation of virtual order into actual order. This happens when an electron occupies a virtual state of an atom or molecule and turns it into an occupied state. In this process a virtual wave function is transformed into the actual wave function of an occupied state. Interestingly, both types of wave functions involved in this process—the virtual and the actual—are the same sort of thing, if I can use the term *thing* in this context: Both of them are lists of numbers.

The two lists of numbers differ in an important point: An actual state can give rise to the emergence of a probability pattern when it is observed repeatedly. An empty state can't do that. Even the visible probability distributions of occupied states aren't visible in a single measurement, but emerge in repeated measurements.

It is really an important puzzle how the mind and its thoughts can affect the material brain and the world. How can something nonenergetic affect energetic processes? It is possible because the first step from the virtual to the actual involves entities of a single type. We can only describe these entities in a mathematical way—that is, as numbers.

WHY VIRTUAL STATES ARE REAL: NONEMPIRICAL DOESN'T MEAN NOT-REAL

In the appendix accompanying this chapter, I have added for you a section entitled "Are Virtual States Real?" It presents additional arguments for the view that virtual states are truly existing entities—that is, physically real elements of the world. They aren't mere mathematical tricks that physicists use, as Niels Bohr thought, to rationalize their experiments. And they aren't so strange, because quantum theory is merely a fragment of a deeper theory that will soon return physics to sanity, as Einstein thought.

In an empirical science the discovery of a nonempirical reality is an embarrassment, like the illegitimate child of a fallen daughter in a pious family. A young scientist walking a nonempirical path may soon find herself fired, where the verdict "You are fired" is often pronounced with a certain feeling of nostalgia and regret that heretics can't be set on real fire anymore. Many pioneers, among them such giants as Niels Bohr and Albert Einstein, constantly reassured their clients that, because they are invisible, wave functions and, particularly, virtual states are epistemological, not ontological entities. That is, they have nothing to do with the nature of reality, but only with our knowledge of reality.

I like to leave no doubt about my own position: The arguments presented in this chapter and in the accompanying appendix leave no choice but to consider the virtual states of material systems as truly existing forms, even though they are invisible. They are invisible, because they are empty of matter and energy. But they are real because they are forms of potentiality that can affect visible phenomena. Specifically, the virtual states can steer the outcome of chemical reactions, and they influence the spectroscopic properties of atoms and molecules. Like the thoughts in your mind, the virtual states in molecules can do such things by expressing a formal order that already exists before it manifests itself in the empirical world. If the virtual states didn't exist as a realm of potentiality in this universe, Parmenides, the sixth-century-BCE Greek philosopher, would be right: Nothing could happen in this world and there could be no change, because all atoms and molecules need virtual states to do anything. If the universe didn't contain a background of potentiality, nothing could happen in it.

At this point you could ask, Does it matter? The virtual states are empty, I can't see them, why should I care? The answer is easy: You bet it matters. There is a realm of reality here that can act on you, even though you can't see it. If its forms are thoughtlike, they will find a way to affect your own thinking. Since this part of the world is a realm of potentiality, you can rest assured that it will find a way to define your personal potential. Once you consider that such a connectedness is possible, you may develop an understanding of yourself unlike any other that you ever had before.

The virtual states in atoms and molecules show that materialism—the doctrine that to be is to be material—is wrong. "Modern atomic theory," Werner Heisenberg writes, is "essentially different from that of antiquity in that it no longer allows any re-interpretation or elaboration to make it fit into a naïve materialistic concept of the universe."

Note his choice of words: Materialism is naïve! The reason for this verdict: "Atoms are no longer material bodies in the proper sense of the word."

Virtual state actualization in quantum jumps can serve as a model for how the visible order in the universe emerges out of a transcendent order. The universe is shot through with virtual states that are waiting for their chances. From the incessant dance at the frontier of the visible world, in and out of the cosmic potentiality, the invitation to you is clear: Reach into the realm of forms, try to play with them, make them a part of your own potential, and join the dance.

What can we know about the nonempirical part of physical reality? That question is important because, after all, we depend on this part of the world, and we ought to find out what we are dealing with. Even though exploring a nonempirical part of the world isn't the task of physics, many quantum physicists in the past century couldn't stay away from that topic. One of the first qualities of the nonempirical world that they discovered was wholeness. Next, we'll take a look at what this concept implies and what it means for us.

Chapter 3

WE ARE ALL CONNECTED: REALITY AS INDIVISIBLE WHOLENESS

"Relativity and quantum theory agree, in that they both imply the need to look on the world as an undivided whole, *in which all parts of the universe, including the observer and his instruments, merge and unite in one totality . . . The new form of insight can perhaps best be called* Undivided Wholeness in Flowing Movement. *This view implies that flow is, in some sense, prior to that of the 'things' that can be seen to form and dissolve in the flow . . . In this flow, mind and matter are not separate substances. Rather, they are different aspects of one whole and unbroken movement."*

—DAVID BOHM

The previous chapters have shown us this much: There is a part of reality that we can't see, but it is real, because it can act on us. We saw that this part of the world doesn't consist of material things but of empty, or virtual, forms that represent a realm of probabilities and potentialities. Our next move will be to take a closer look at this transcendent basis of the world. Specifically, we'll consider some hints that the transcendent forms are hanging together; that is, they are forming a wholeness. If it turns out that the background of the universe is an indivisible wholeness, this will be of the greatest significance for us, because it will mean that we belong to the wholeness and that the cosmic wholeness is active in us. Specifically, this would mean that the inner potential in you is cosmic. When you actualize your potential, the cosmic wholeness is actualizing in you. You could say that you are an embodiment of the cosmic potentiality. This leads to a view of our human nature that is radically different from everything that science has ever taught.

The concept of wholeness means that an invisible realm of reality, in which all things are interconnected, is underlying the visible world. If that is so, everything comes out of this wholeness. Everything is guided by this wholeness. Everything in the universe is connected with everything else through this wholeness. There are indications that such connections exist, even though the world seems to consist of separate things and people. However, the things of this world are disconnected only in their empirical existence, but connected in their nonempirical roots. Since the nonempirical roots aren't visible, the assumption of the wholeness has led to much controversy. Some quantum physicists accept this aspect of the world while others reject it. Personally, I find the evidence convincing. I will describe it to you, so that you can form your own opinion. One way to deal with the uncertainty of a question is to live the question. In this case, start out by telling yourself that you are a meaningless, worthless, and accidental material structure, alien to the world and isolated from everything else. Act in the spirit of this principle and live with it for a while, and then see where that will take you. Then, at the right moment, make the switch to the view of wholeness: Live within the wholeness as though it existed, taking the connectedness with the universe as a fact, and see where that will take you. Living contradictory views of reality can give us a feeling, if not proof of, which one is more likely to be true.

David Bohm was one of the great pioneers of quantum physics and an inspired man. During his career he developed the view that the ultimate reality is an unknowable wholeness. David Bohm's choice of words is different from mine, but his views are similar. You could take the term "transempirical," for example, to describe the same aspect of the world as "unknowable." Inspired by holography, a special technique of photography, Bohm proposed that the "total order" of the world "is contained, in some implicit sense, in each region of space and time." By this Bohm meant that each region of the world contains the "total structure 'enfolded' within it." The tiniest parts of the universe, your brain and the atoms in it, contain the entire order of the world.

If all appearances of the world contain the entire order of the world, the flow of the waves in interference phenomena and their connectedness are of general significance. They inspired Bohm to think not only that reality is an "unbroken and undivided totality," but also that the "undivided wholeness is flowing movement." Out of the constantly changing flux, he said, we can lift up and "relevate"—or make relevant—the elements of our experience of the world, including life, mind, and matter. These elements don't exist in the wholeness, Bohm thought, in the way in which they appear to us. In the flow of the wholeness, "mind and matter are not separate substances. Rather, they are different aspects of one whole and unbroken movement." Specifically, the elementary particles of our experience aren't eternally enduring things; we have to look at them, as Bohm described it, like vortices in a flowing river.

Bohm's reasoning is a good way to start the discussion. Underlying the visible world is the realm of potentiality, and its nature is that of an indivisible wholeness—the One. We have to think that the One is guiding not only the physical processes of the world, but also the processes of our mind. Thus, the nature of the world is an issue that is important in many ways, beyond the limited concerns of physicists. We should know about these things because they have consequences for how we should live. Since we live in this world, we should understand its order because we have to live in accordance with it. We can't live outside of it. For example, you can't defy gravity, jump off the roof of your home, and think you will live happily ever after. In the same way, you can't jump out of the wholeness and expect that you can get away with it. Just think how you would interact differently with people in a holistic universe in which you are connected to the rest of humanity than in a world of isolated material things and beasts of prey, where the Darwinian virtues of aggression and selfishness are adequate. This is just one example of why understanding the wholeness is so important for our lives. Many other examples can be found and will be described later on in this book.

Connectedness and wholeness also play an important role in the practice of meditation. As Deepak Chopra explains, the purpose of meditation is "to tune in." Between any two thoughts in our mind, there is a gap that is a window, or corridor, to the "infinite mind," which many people call *God*. According to ancient wisdom traditions, Chopra points out, this gap is a field of "infinity potentiality" in which "everything is connected to everything else."

The Opposition:
The Reductionist View of the World

Scientists often take pride in the fact that their methods of studying the world are reductionist. Reductionism is any method of studying complicated systems by taking them apart, trying to understand the essence of a system by the properties of its parts. For example, you might try to understand how a clock works by taking out all its cogs and gears. The idea is to simplify a complex principle by reducing it to simpler, more fundamental ones.

Apart from being a procedure, reductionism is also a philosophy, a metaphysic of what reality is like, and it is even a moral system, exhorting scientists as to how they ought to think. Steven Weinberg is a prominent physicist and author of several books on how physics works, among them *Dreams of a Final Theory.* To Weinberg, reductionism isn't only a prescription for what scientists should do. It isn't "a guideline for research programs, but an attitude toward nature itself . . . a statement of the order of nature."

In this definition, reductionism is the doctrine that being reducible is a property of nature—that the essence of reality is found in its parts. For example, you can understand the properties of materials by studying the properties of their molecules. You can understand molecules by understanding the properties of atoms, atoms by understanding even smaller elementary particles, and so on. The question is how deep you have to dig to discover the roots of all things to which everything can be reduced. But the question is also what you may lose when you take a system apart.

In their studies of life, biochemists, for example, divide living cells into smaller and smaller parts until they are dead and all life has been lost. If you took all the right material components that could be found in a living cell and mix them together, would the mix begin to live? How come nobody has succeeded in doing that yet?

In the same way, you can take your brain apart until all thoughts are lost. Will we be able someday to take the right number of neurons and screw them together in such a way to start a thinking brain?

These questions show that reductionism as a metaphysic of what reality is like has its problems. It is very good at finding out how the fragments of a system work, but it is always in danger of missing the essence of what it studies. Because of such considerations, more and more scientists are beginning to suspect that not only biology and brain science, but all disciplines of reductionist science have this problem. Specifically, the intuition is growing among some scientists that the universe is a coherent wholeness, like an organism, whose life is destroyed when it is taken apart. This is the conflict between the reductionist view and the holistic view of the world.

Weinberg is aware of this polarity, and he takes pleasure in rubbing it in. "At the other end of the spectrum are the opponents of reductionism who are appalled by what they feel to be the bleakness of modern science," he writes. By the supposed "bleakness" he means the reductionist view. "To whatever extent they and their world can be reduced to a matter of particles or fields and their interactions," he writes, "they feel diminished by that knowledge."

Here I have to confess that I don't quite understand the problem. Steven Weinberg is a brilliant scientist, but why should I feel diminished by any true "knowledge" of what the world and I are about? I think that the problem isn't that people feel diminished. Rather, the problem is that it is unlikely that our hopes, loves, and feelings are reducible to the empirical properties of a few material particles that are nothing but lumps of matter or energy. The problem isn't the question of human dignity, but credibility. It is difficult to see how you can disassemble my body into its parts and find out who I am. Above all, nobody has yet been able to take a bunch of elementary units of matter and make a living thing out of them.

At this point, you could accuse me of being inconsistent. In the previous chapters we've been generating big ideas about the nature of the universe and life by looking at ETs, elementary things or thoughts—breaking the system down to some of its smallest parts. How has that approach not been reductionist in Weinberg's sense? In other words: How can I claim that biologists are missing the essence of life when they take it apart, and then I turn around and do the same sort of thing; that is, I glean the essence of the world from its smallest parts?

The answer lies in the way in which reductionists understand the parts of a system. Their parts are material, they are independent of one another, and what you see is what you get. That is, the particles that reductionists study don't contain an invisible order. They are enduring bits of matter, and they don't make transitions into a nonmaterial existence in a transcendent part of reality. And, most important, they aren't connected with one another in a holistic world. They are only matter or energy or empirical fields, but not fields of thoughts or empty forms.

The following quote will show that this is the kind of reductionism that Weinberg has in mind. He has little respect for those who dare to oppose his views; and, in a delightful way, he loves to stir the pot. "At its nuttiest extreme," he writes, "are those with holistics in their heads, those whose reaction to reductionism takes the form of a belief in psychic energies, life forces that cannot be described in terms of the ordinary laws of inanimate nature."

For all these nuts Weinberg has this message: "The reductionist worldview *is* chilling and impersonal. It has to be accepted as it is, not because we like it, but because that is the way the world works."

Here, again, I have to ask: If this is how the world ticks and if this is our nature, why should it be chilling? And why are those scientists who are asking questions nutty? You could also put the question by asking why true science has to be reductionist. Asking questions is what science is about. Moreover, stating that a transcendent order and potentiality exist in atoms and molecules is expressing a scientific fact. Why should I not be allowed to think that my own transcendent nature and potential are part of a greater whole?

My method is as reductionist as Weinberg's method. But if you want to be a reductionist, you have to go to the bottom of things, mustn't stop at an intermediate level, like the level of material particles. Going to the bottom of things means going to the level of ETs, at which they turn into nonempirical forms.

Nothing that I am saying is in conflict with the "ordinary laws of inanimate nature." The difference is that I don't disregard the visible hints of a transcendent reality that the traditional reductionists try to deny. The conflict isn't between reductionism and holism. Rather, it is between a complete view of the world and a selective view. The problem with the "ordinary laws of inanimate nature" is the fact that they are incomplete, because they are the laws of the visible surface of the empirical world.

Traditional reductionism is empiricist, materialist, and separatist reductionism. It is, of course, not the only opposition to my views. At the opposite end you will find the religious fundamentalists, who claim a monopoly on truth, because their views of the world are based on sacred archaic texts that mustn't be questioned. But, in descriptions of the world, nothing is sacred. The only thing that counts is, whether a given description is meaningful or not.

How the Concept of the Wholeness Emerges in Quantum Coherence

In quantum theory the concept of *coherence* denotes the ability of a single ET to *interfere*. That ability is apparent in interference phenomena. *Decoherence* denotes all processes that diminish or destroy the coherence of a system. For a particle in the state of a potentiality wave, for example, decoherence happens when it gets entangled with other ETs in its environment.

The interference phenomena observed when single electrons pass through two slits are typical signs of coherence. The word *coherent* derives from a Latin word that means "hanging together." In single-electron interference phenomena, for example, you can think that the two states of an ET—that is, passing through slit number 1 and passing through slit number 2—are hanging together, in the sense that their simultaneous presence forms a single state.

Coherence is a trademark of quantum physics that has no equivalent in Newton's physics. For quantum systems the hanging together can be so intimate that a wholeness results in which the participating states or systems lose their identity.

Quantum coherence is one of the aspects of the quantum world that led physicist David Bohm to the view that wholeness is a general property of reality, meaning that an "implicate order" exists at the cosmic level—a level in which every particle "is in instant contact with every other particle in the universe," as Nick Herbert describes in his popular book *Quantum Reality*. Cosmic interconnectedness can be thought to be mediated by a cosmic realm of potentiality in which all things and people are connected. You can think that this is the realm in which elementary particles dissolve when they become potentiality waves. When an ET enters the realm of potentiality, it loses its identity and becomes a cosmic event.

Many other physicists have expressed suggestions of wholeness. Hans-Peter Dürr and Marianne Oesterreicher, for example, write: "Quantum physics has revealed a connectedness of everything with everything that eludes any manipulative intrusion." Dürr also compares the cosmic potentiality to an ocean, and then the concept of spirit comes into play. "When the ocean is completely quiet," he says, it means that "Spirit has not yet expressed itself."

That wholeness is a property of the entire universe can't be observed and can't be demonstrated. The idea appeared in the minds of physicists who were open to it after they observed specific examples of coherence. It is like when your parents first took you to the beach and let you swim in the ocean: The experience made you believe that all oceans are wet everywhere, even in their distant parts that you have never visited.

How the Concept of Wholeness Arises from Potentiality

Different physicists were led in different ways to the assumption of wholeness. For some it was the experience of quantum coherence. For others it was the encounter with the potentiality in physical reality. Hans-Peter Dürr, for example, was led to the concept of wholeness by realizing that "reality reveals itself primarily as nothing but potentiality." Personally, I find that the totality of all the various arguments has a cooperative effect that builds up an increasingly convincing case for a cosmic wholeness. No measurement of the world and no experiment can prove the wholeness of the universe, but it is a plausible hypothesis—and more plausible, at that, than the opposite claim that a holistic nature of the universe is impossible.

We have seen how ETs, when they leave the particle state and enter the realm of potentiality, develop a probability of presence that is increasingly spread out in space. We can't look inside the realm of potentiality and don't really know what is going on in there. We can see only how it interacts with the empirical world. We don't even know that our notion of space and time applies to this realm. But whatever the state of potentiality of an ET might be, in the empirical world there is a spreading of the probability of its presence. This shows that the forms in the realm of potentiality are spreading out in the sense that their interactions with the empirical world don't stay localized at a single point. Thus, it is reasonable to think that the forms of different ETs in the realm of potentiality will superimpose and interact with one another, like all waves do. In this way the aspect of wholeness suggests itself quite naturally in connection with a realm of potentiality. As Dürr describes it: "Potentiality appears as the One—better yet, as the Not-Twofold—which cannot be dissected or separated into parts."

In considering such statements, we must always keep in mind the differences between the facts of quantum science and the assumptions regarding the nature of the world that are suggested by the facts. Many properties of the quantum world are facts, even those that violate common sense. The invisible wave states of ETs are such a fact. The existence of a realm of potentiality, too, is such a fact, and so is the existence of a nonempirical realm of reality. In contrast, suggestions regarding the wholeness of the universe can't be considered factual in the same way. Of course, I believe that they are meaningful, and that's why I am describing these aspects to you.

The images in the realm of potentiality exist whether they actualize or not. Since they are transempirical, we can't "know" what they are like. But since they display the properties of waves, it is reasonable to think that they act like all waves do: They spread out, superimpose, form a coherent system, and play with each other. If we accept this view, the notion arises that the background of the visible world is an ocean of potentiality whose waves form the wholeness of the One, in which all things and people are interconnected.

We are surrounded by countless things that seem separate and independent of one another. You can walk through your living room without having to drag the furniture along with you. Similarly, you know a lot of people, and you don't feel connected with many of them. So how can it be possible that a realm of wholeness exists in the universe, in which all things and people are connected? The answer is that the connection doesn't exist in the visible world, but in the transcendent roots of things. Similarly, your connection with other people doesn't have to be a part of your consciousness, but it can exist in an unconscious realm of forms, with which we are all connected.

Remember the white caps on the waves of an agitated ocean? Underlying the visible world is an invisible field, which is best described as an invisible ocean whose waves swing up to visible caps. The caps can be electrons or protons or atoms or people or flowers. These visible caps look like isolated objects but are all interconnected in their transcendent roots.

I have said that potentiality waves are spread out in space, but we don't even know that the nonempirical realm of reality has the same space and time as our empirical world. This may be the reason why Schrödinger's waves have the meaning of probabilities, and chance plays a role in the quantum phenomena: Chance and probability are all that we can know about how the ETs in the realm of potentiality interact with the empirical world.

Giordano Bruno, the sixteenth-century Dominican friar who was burned at the stake in 1600, developed a view of a realm of potentiality in the world that allows an easy connection with the aspect of wholeness. The structure of reality, Bruno wrote, is "coincidence of matter and form, potency and act, so that being, logically divided into what it is and what it can be, is physically undivided and one." Thus, in the same way as in quantum reality, Bruno's world is made up of what is and the potential of what can be.

There is an unexpected aspect of Bruno's statement, in that it seems to suggest that a reality that is structured into "matter and form, and potency and act" is necessarily "physically undivided and one," as though the aspect of wholeness is a necessary condition of every system whose visible surface is an emanation out of a realm of potentiality.

We can try to understand Bruno's argument in the following way: If the physical reality consists of a nonempirical realm of potentiality and an empirical realm of material things, and if the realm of potentiality wasn't undivided and one but instead consisted of numerous disconnected and independent fragments, the uncoordinated actualizations out of the fragments of potentiality would lead to chaos in the empirical world, and would make a coherent understanding of it impossible. Since such chaos doesn't exist, the aspect of the wholeness of the transcendent realm of reality is a logical necessity.

From the unity of our thinking Bruno postulated the unity of the world: "For you must know that it is by one and the same ladder that nature descends to the production of things and the intellect ascends to the knowledge of them; and that the one and the other proceeds from unity and returns to it."

"The universe is all in one: infinite, unmoved, possible, real, form, soul, matter, cause, essence, purpose," Johannes Hirschberger writes about Bruno's philosophy. Even the individual "is nothing in itself, but only a modification of that which is one and all." It follows that, if you are searching for happiness outside of this cosmic connection, you won't find it.

In this way the concept of wholeness suggests itself quite naturally: Like the way water waves in an ocean hang together, the waves in the ocean of potentiality hang together and our thoughts hang together and form a coherent whole. In a more technical language, physicists speak of the "nonclassical coherence" of the states in the realm of potentiality. This is a kind of coherence that the objects of classical physics aren't capable of. From the appearances of such a coherence in special instances developed the view that it is a general property of the world, so that physical reality is ultimately an indivisible wholeness—the One—in which all minds are interconnected.

Since the forms of the realm of potentiality are nonmaterial and invisible, they are more thoughtlike than thinglike. They are patterns of information, like thoughts. They are potentiality, like thoughts. For such reasons Hans-Peter Dürr made the connection between the ocean of the cosmic potentiality and the emergence of "spirit" in the world.

It is often said that all things are connected in one giant energy field that fills the universe. Maybe there is such a field, but the realm of potentiality isn't an energy field. Its forms aren't energy. Like ideas, they are potentiality. Their power over the world of matter is like the power of the ideas that you feel in you.

How the Concept of the Wholeness Arises in the Nonlocal Nature of Quantum Systems

The concept of nonlocality plays an important role in quantum physics (see the appendix to chapter 1 on page 242). When an electron passes a double slit and approaches a detector, immediately prior to being detected it has a certain freedom of choice of where it will appear. It will always appear in just one of the channels of a detector, but before its appearance each channel has a true chance to catch it. When the electron is actually observed at a specific point, instantly the probability to find it drops to zero everywhere else.

This is an example of a nonlocal event: Something you do at one point in space—in this case, catching an electron—has an instantaneous effect somewhere else—in this case, changing a local potentiality in all channels of the detector. Nonlocality means that influences exist whose effects propagate through space at a speed faster than light.

You can also describe nonlocality in the following way: We speak of nonlocality when an observation made by an observer in one laboratory has an instantaneous effect on the outcome of an experiment performed in another laboratory a long distance away.

At this point you may ask yourself why nonlocality is so special. The answer is that, in the visible world no signal—no influence—can move faster than light. The speed of light is the upper limit at which anything can move in this world. It is a true limitation of our physical existence. When the speed limit on a highway is 75 mph, you can go 85 mph and get away with it, if no police officer sees you. But, when the absolute speed limit for anything in this universe is the speed of light, you can't go the speed of light plus 10 mph and think you will get away with it. That limit is absolute! Nothing can surpass it.

Thus, nonlocality shouldn't exist because it involves instantaneous transmissions of signals from one point in space to another. But signals can't travel faster than light. This is the reason, for example, why there is always an unavoidable delay between a question and answer when you talk to an astronaut out in space. In contrast, in the quantum world influences can act nonlocally, that is, instantaneously over arbitrarily long distances.

If you want to further develop an intuitive understanding of the nature of nonlocality, you can think of some mental processes, such as extrasensory perception or telepathy, that typically appear with aspects of nonlocality. Telepathy involves the transfer of thoughts or emotions from one person to another, through long distances and without any delay. Twin telepathy is a fairly well-established phenomenon. There are many examples that show that twins are connected in a special way. When one of them is in distress, the other one will be aware of it instantly, no matter how far apart they are.

When two particles interact, they can form a special connection in a state called a *twin state,* like twins do. In this state they are so intimately entangled with each other that they "respond together to further interactions" with their environment, as Menas Kafatos and Robert Nadeau describe this phenomenon. As it turns out, such states are characterized by an amazing property: They continue to exist when the two entangled particles move away from each other to different parts of the world. In that case, an observer interacting with one of the twin particles in her laboratory will instantly affect the other one, no matter how far away it is.

Experiments of this kind are often referred to as *Bell-type experiments* in honor of John Stewart Bell, the Irish physicist who invented them. When these experiments were first performed, physicists were in shock, because nonlocal phenomena violate a sacred dogma of physics. In his book *In Search of Reality,* physicist Bernard d'Espagnat points out how such phenomena force us to think that two twinned particles "even if they occupy regions of

space that are extremely far from one another—are not truly separated." He calls this property of ETs in such coherent states their
nonseparability.

Strictly speaking, nonseparability was first discovered in experiments involving special types of particles and special conditions. For
these particles nonseparability denotes a special state of wholeness.
To what extent it is permissible to assume that such a state exists for
all particles in the universe, which at one time interacted and then
moved to different galaxies—that is an interesting question. There
are no experiments that can be performed that will take a look
at all the particles in the universe and see whether they are connected. But, d'Espagnat points out, the same theory that correctly
predicted the observed cases of nonseparability also makes it possible to predict that all the particles in the universe are nonlocally
interconnected. Thus it can lead us to the conclusion "that nonseparability is a most general feature" of the world, as d'Espagnat
described it. If nonseparability is, indeed, a general feature of the
universe, it follows that the universe is a holistic system.

Kafatos and Nadeau pointed out that observations of special
cases of nonlocality may allow us to "infer" that reality is an undivided wholeness, but don't "prove" that. A part of this difficulty is
due to the fact that we can't know what is going on between two
nonlocal connected events in Bell-type experiments. As Kafatos
and Nadeau describe it, we can't measure "the reality that exists
between the two points." It is as though the twin particles exist in
a different kind of reality, outside of our space and time. When
something happens at two different places in our world, it's happening at a single point in their world, as though the space between
the two particles-in-themselves doesn't exist.

Needless to say, in our current life-form we can't ever know
what is going on in a reality outside of our space and time, if such a
reality, indeed, exists. "All that we can say about this reality," Kafatos and Nadeau write, "is that it appears to be an indivisible whole

whose existence is 'inferred' where there is an interaction with an observer, or with instruments of observation, and that it also appears to exist outside of or beyond space-time."

Nonlocality is yet another sign that a realm of reality exists that is transempirical. We have to think that the nonempirical reality that Kafatos and Nadeau describe is the same as the nonempirical realm of potentiality that we discussed in the previous chapters.

The nonlocality or nonseparability of the universe is another nonempirical aspect. That two particles can be nonlocally connected is an experimentally tested statement. That the whole universe is a nonseparable system is an entirely different case: It is unobservable. At the frontier of reality we see the limits of our ability to obtain information on the world by observing it. We can only look inside and try to find likely intuitions or expectations. The art is to separate the intuitions from the illusions.

Even though they are cautious, Kafatos and Nadeau leave no doubts about their expectations. They point out that literally all material particles in our environment, including the atoms in your body and mine, are made up of smaller quantum particles that at one time in the history of the universe interacted with one another and formed nonseparable states. From this it is possible to infer that nonseparability is "a fundamental property of the entire universe," or a "factual condition" of reality, as Kafatos and Nadeau put it. All of reality is one, and you and I belong to it.

THE COSMIC POTENTIAL IN YOU

At this point you could ask what all of this means for us as individuals and as a society. You are living in your community. You interact with the people of your choice, keeping others out of your life—why should you care about the interconnectedness and the wholeness of all, if that isn't even an observable aspect of the world?

The answer is that it doesn't matter whether anything is visible or empirical or not. What matters is, is it real? If it is real, it can affect you. It seems reasonable to expect that the structure of living beings reflects the structure of the universe. So you should know about it. What if, for example, you need the contact with the cosmic potentiality to develop your own potential to a level where you can lead a happy life? But how could you get in touch, if the cosmic potentiality weren't connected with you? Nothing else is meant by the principle of wholeness: It is the state of being connected and the ability to consciously connect.

Physicists discovered the realm of potentiality in the physical reality at about the same time psychologists discovered a corresponding realm in the human psyche: the inner potential in us. We owe this insight to Abraham Maslow and his "humanistic psychology."

As Abraham Maslow described it, our life is driven by a system of basic needs that form a hierarchical structure in our mind, which he compared to the steps of a "pyramid." This structure exists in us like a system program in a computer, and it drives our actions. At the most basic level of our needs we find the physiological needs. We all have to eat and drink and sleep; those are the basics. At higher levels we find, in ascending order, needs of safety, of love and belonging, and, finally, of self-esteem. All these needs form a hierarchy in the sense that each level is active only when the lower levels are satisfied. When you haven't eaten in days, for example, and you are starving to death, you won't be troubled very much by your self-esteem when you stumble over a bag of food in the street and you have neither fork nor knife to eat it with etiquette.

It is an important aspect of the pyramid of needs that, when you have taken care of all of them, you might still be restless or discontented with your life. That kind of restlessness can evolve, for example, in a person who has focused his life on serving the needs of his body, while disregarding his spiritual needs. This restlessness has to do with the potential in us that Maslow discovered.

We all carry in us an inner potential that needs to be actualized. I am convinced that this potential is an expression of the cosmic potentiality that tries to actualize in us. We won't find peace of mind when we neglect our inner potential, whatever it may be. "A musician must make music," Maslow wrote, "an artist must paint, a poet must write, if he is to be ultimately happy. What a man can be, he must be."

Maslow called the need of actualizing our inner potential the need of "self-actualization." This term, as he explains it, "refers to the desire for self-fulfillment, namely to the tendency for [a person] to become actualized in what he is potentially. This tendency might be phrased as the desire to become more and more what one is, to become everything that one is capable of becoming." I am sure that the similarity of the process that Maslow described with the process by which the empirical world is actualized out of the cosmic potentiality is clear.

If you wonder how Maslow arrived at his theory—apart from the fact that it was a part of his potential—you can look at your own life. Have you ever felt some restlessness, even though all aspects of your life are under control and they have turned out as successful as you could ever have hoped for? You have a nice family, wonderful friends, no financial problems, and, yet, something is missing? If that is the case, what is missing is the expression of this mysterious potential in you. I call it mysterious, because nobody knows where it is coming from. It's usually said that it comes out of our genes. But what does that mean? Where did our genes get it from?

The suggestion that I like to make is this: We have an inner potential because the universe has an inner realm of potentiality. The inner need in you is a cosmic need, and this is where the wholeness of it all is so important: The inner urge can be in you because the nature of reality is that of a wholeness; that is, it can be in touch with you. Because the universe is an indivisible wholeness, the potential in you is cosmic. Does the fact that its actualization is felt as a personal need mean that the cosmic potentiality is under pressure to actualize in the empirical world?

Since actualizing your potential is a mental process, you can ask at this point how in the world the physical universe can be involved with it. Do we have to think that, in its cosmic potentiality, the universe reveals mindlike properties? Does this mean that when you actualize your inner potential a cosmic spirit is acting in you? What a stunning turn of events that would be—bringing a cosmic spirit into play!

With that our next task presents itself to us. Without intending to do so, the discussion of the quantum properties of the world has led us to questions of psychology. That connection must be explored and the question is this: Are the forms in the cosmic realm of potentiality something mindlike or thoughtlike? If so, does that mean that the universe has mindlike properties, or that consciousness is a cosmic property?

Some of the pioneers of quantum physics have, indeed, argued that point, and I believe that they are right.

Chapter 4

CONSCIOUSNESS:
A COSMIC PROPERTY

"The teaching of Sri Aurobindo starts from that of the ancient sages of India: that behind the appearances of the universe there is the reality of a being and consciousness, a self of all things, one and eternal. All beings are united in that one self and spirit but divided by a certain separativity of consciousness, an ignorance of their true self and reality in the mind, life, and body . . . Sri Aurobindo's teaching states that this one being and consciousness is involved here in matter. Evolution is the process by which it liberates itself; consciousness appears in what seems to be inconscient, and once having appeared is self-impelled to grow higher and higher and at the same time to enlarge and develop toward a greater and greater perfection."

—SRI AUROBINDO

That the universe contains a realm of potentiality is about as factual as any factual statement we can make about the world. Potentiality is everywhere: We find it in the probability waves into which the ETs dissolve when they are left alone, and we find it in the virtual states of atoms and molecules. That human beings, too, have an inner potential is a fact of psychology. Since the potential in us involves our mind and, with it, our consciousness, this brings up the question whether the cosmic potentiality is a sign that the background of the universe is also mindlike and whether consciousness is a cosmic property. The question implies a congruence of the physical and the mental that many people will find unacceptable. When this idea first came to my mind and I decided to discuss it in my first book, *In Search of Divine Reality,* I was shocked when I saw what I had put on paper, and I wondered whether I would dare to put it in print. As it turns out, many physicists in the twentieth century, and great pioneers among them, expressed such views a long time before I did.

If the background of the universe is mindlike, it is possible that consciousness is a cosmic property. I will refer to this as the *cosmic consciousness.*

How the Cosmic Consciousness Reveals Itself in Subtle Suggestions

The idea that some form of consciousness is active at the cosmic level didn't appear with the quantum phenomena in one flash or thunderbolt. There was no knocking on someone's door and a friendly voice that said, "May I introduce myself? I am your Cosmic Consciousness." The idea is so alien to science that there were no sudden revelations of any kind—no experiments or peak experiences. The idea suggested itself gently in subtle ways, at first in small steps and with great hesitation, but then with increasing momentum until it asserted itself with confidence.

To get an impression of the process, consider the many situations in which the concept of information arises in quantum physics as a causal principle. When electrons enter an atom, for example, they are sensitive to information on which of the states of the atom are occupied, and which ones are empty. The ability of quantum states in atoms and molecules to accept electrons is limited. When a given state contains two electrons, it is full. How do electrons entering an atom obtain information that tells them which of the invisible forms are taken and which are still available? Similarly, consider the concept of probability waves that is so important for quantum physics. Probabilities are ratios of numbers—patterns of information. Information isn't matterlike; it's mindlike. Whenever someone talks of information, you are automatically prompted to search for a mind in which that information makes sense.

Take, as another example, the guidance of chemical reactions by nonmaterial forms. As we have seen, the invisible orbital structure of oxygen molecules is the basis of their ability to be used in our metabolism. You couldn't digest your food if it wasn't for the invisible degenerate states of oxygen molecules. All chemical reactions are guided by such forms, which are ultimately not matter and not energy, but just patterns of information. The only other thing that we know that reacts to information is a mind, and it isn't really a thing. Thus, at the foundation of ordinary things we find ingredients with mindlike properties.

Similar thoughts come to mind in connection with the concept of potentiality waves. Waves in the realm of potentiality are nonmaterial. They are more thoughtlike than thinglike, because thoughts have the nature of potentiality. If thoughts belong to a realm of potentiality, does it mean the cosmic potentiality is a realm of thoughts? So, here we go again! In whose mind do these cosmic thoughts exist?

At this point, these are just questions and suggestions. They explore possibilities. But when more and more questions of this type come up, all of a sudden possibility turns into actuality, and suspicions turn into facts.

If you are like most people, you will probably not spend much time thinking about the nature of your thoughts. You just think! Whatever thoughts are, you take them for granted. So if someone comes up with the idea that the nature of the thoughts in you is that of a potentiality, this might amaze you. But, if you think about it, a thought can exist in your mind a long time before it comes out as a spoken word or a written text. When it is in your mind it is real; it isn't nothing, but it isn't observable. It isn't a part of the empirical world, but it has the potential to be expressed in it, like the virtual states of quantum systems can be expressed in material forms.

The point I am making is this: Thoughts are a form of potentiality. This doesn't prove that other forms of potentiality are also thoughtlike, but it suggests it is worthwhile to consider.

It is interesting that, already in the fourth century CE, Augustine of Hippo wondered about the potentiality of thoughts. Speaking to some of his disciples, it occurred to him how strange it was that he had to put his thoughts into some language, such as Latin or Greek, to speak about them. "But the word is neither in Latin in me nor in Greek," he said. "It is entirely beyond any language, what is in my heart." Being beyond language but being able to be expressed in a language is the potentiality of thoughts.

So, here you have it, one more time: Thoughts in you are potentials that exist in your mind. Are the forms of the cosmic potentiality also thoughts that exist in some mind? If they are thoughts, who is waiting to speak them?

In considering these aspects of thoughts, Augustine discovered a hidden principle of great general significance, which we could call the *giving-not-losing principle* of potentiality. The principle says that a realm of potentiality doesn't lose a form when it is actualized. For example, when you share a thought with a friend, you give it away, but you don't lose it. "All right," Augustine explained to his students, "you have heard what is in my heart, so now it is also in yours. It is in my heart and in your heart. You have started to own it, and I have not lost it."

You will find that the same principle applies to the cosmic potentiality. The forms that actualize in the empirical world continue to exist in the cosmic potentiality. If that weren't the case, we would be in bad shape! Where would this world be if the actualization of a single hydrogen atom would erase its form from the cosmic potentiality? Fortunately, the cosmic potentiality is like a telephone book. Each number is listed once, but you can call it as often as you like. In the same way, a library can have a single copy of a book, but it can loan it out to many people.

At this point you may realize with some astonishment that the giving-not-losing principle of operation turns the cosmic potentiality into a memory field. It remembers the forms that it has dismissed into actuality. But memory has something to do with mind, and a cosmic memory with a cosmic mind. It is the craziest thing, but wherever we turn in this quantum world, aspects of mind appear at the cosmic level, insisting that they are real.

You might object that a computer has a memory but no mind with its consciousness. That is true, but the memory of a computer wouldn't exist without the human mind. So instead of solving the problem, you have aggravated it. If the memory of a computer wouldn't exist without the human mind, does that mean that the human mind wouldn't exist without a cosmic mind? And who or what might that be? I am not asking this question to give it any answer, but to show you how the idea of a cosmic consciousness is constantly imposing itself upon us when we consider the quantum world, whether we like it or not.

Augustine wouldn't be the saint that some people think he was, if he hadn't found a deeper lesson in the giving-not-losing potentiality of thoughts: "Like my word assumed a language by which it was heard," he concluded, "so the word of God assumed flesh, by which it was seen."

I don't know whether you realize this, but this is your own story: At one time you were a word in the cosmic potentiality. And then, bingo, before you knew it, "your number came up," as Jacques Monod, the prominent French biologist and Nobel laureate, described it. Your word was spoken and dismissed into the empirical world. But the cosmic potentiality didn't lose you. You are still in it and you will be in it after you die. Maybe we are so fond of some people because they are actualizations of the same forms that actualized in us. Interestingly, the same thought is found in the Bible: "Before I formed you in the womb I knew you."

Sir Arthur Stanley Eddington, prominent astrophysicist of the last century, developed a strong view of a cosmic consciousness. One of the phenomena that inspired his view was the spontaneity of quantum processes. Spontaneity is the absence of causality. When you do something spontaneously, you may do it for no good reason. A thought comes to your mind, you like it, and all of a sudden you are painting the doorway to your apartment pink. Just like that! When a molecule changes its quantum state, it does the same sort of thing: makes a spontaneous jump; no cause. A self-conscious mind is the only other thing that we know that can act in this way. Thus, again, there is the suggestion of a presence of mind.

A serious consequence of the absence of causality in the actions of quantum systems is, as Eddington described it, "that it leaves us with no clear distinction between the natural and the supernatural." Thus, at the quantum level of reality, the borderline is blurred between the natural and the supernatural, between the physical and the metaphysical. So why not between the mental and the material? All these attributes make sense when you are talking about the nonempirical realm of reality.

We can also describe the situation in this way: At the level of elementary particles, idealike forms become matterlike. In biblical terms, the word is becoming flesh. All actualizations are materializations of thoughts. King Midas had the gift of turning things into gold merely by touching them. We can turn potentiality forms into matter by watching them. It was part of the classical mind-set to think that, if one proceeds to smaller and smaller parts of space, at its frontiers, physical reality turns into nothing, or empty space. Now we are finding out that the transition of the visible world isn't into nothing, but into a transcendent order that is mindlike. The physical realm of the world is based on a metaphysical realm. And, even more shocking, the metaphysical is primary and the physical is secondary. Material things border on the mental; they even dissolve into something mental. The basis of the world is a field of information and a storehouse of information. Nothing forbids us to believe that the background of reality is mindlike.

All these considerations are possible, but not necessary. The importance of information in the universe, the rule of nonmaterial probability waves as a basis of reality, the spontaneity of quantum events: All of these phenomena and concepts have mindlike aspects. They don't prove a thing about a cosmic consciousness, but they suggest a lot. This situation is symptomatic. There are no archaic threats: The view from science isn't dogmatic. It offers possibilities and opportunities.

At first the physicists shrugged their shoulders when such aspects of reality became apparent, and the general reaction was that these suggestions didn't mean a thing. Then, step after step, some of them began to wonder whether this really did mean something. And then, all of a sudden, the idea asserted itself: The many hints of consciousness at the cosmic level taken together had to mean that the background of the universe is mindlike.

As time went on, many physicists got used to the possibility that consciousness is a cosmic principle. In the books of physicists Hans-Peter Dürr and Hans-Jürgen Fischbeck, for example, the concept of a cosmic consciousness appears as a given and is accepted without much debate.

In case of doubt, consider the opposite. And then decide for yourself what the most likely option is.

How the Cosmic Consciousness Reveals Itself in the Thoughtlike Background of Atoms

In the 1930s, Sir Arthur Stanley Eddington was the first physicist who systematically pursued aspects of spirit in the quantum phenomena. "The universe is of the nature of 'a thought or sensation in a universal Mind,'" he wrote, shocking the scientific establishment and upsetting many of his colleagues. But Eddington was a courageous man, and he arrived at his conclusions in a logically acceptable way.

Measurements in physics are meaningful, Eddington pointed out, because the measuring instruments are connected with a somehow meaningful background of the things that we measure. For example, when you follow the movements of a light dot in the sky at night, your observations make sense because you know the background that contains the planets revolving around the sun. The problem with our measurements of atoms, Eddington said, is that the background isn't known. Why isn't it known? Because it consists of nonempirical forms. "Now we realize," Eddington writes, "that science has nothing to say as to the intrinsic nature of the atom."

As physics, so neurology. In physics nothing tells us what is going on behind the visible surface of an atom. In neurology, measurements won't tell us what is going on behind the surface of a brain. However, in contrast to atoms, the person behind the brain can talk to us. That person can tell us that, behind the visible surface of the neural networks, there is a mind. Atoms don't talk, but we can put two and two together. When we look at a brain, we see the electric activities of its neuronal networks, and we know that there is a mind behind them. So, Eddington asked, why not link the two together, the atomic and the mental? Since we have to find something to which we attach the measurement of an atom, "why not then attach it to something of spiritual nature of which a prominent characteristic is thought." No question mark, period! And, he continues: "It seems rather silly to prefer to attach it to something of a so-called 'concrete' nature inconsistent with thought, and then to wonder where the thought comes from."

The last part of this statement is remarkable! It can be understood to imply that we can have thoughts in our mind because something akin to thought already exists in the background of atoms. With this, Eddington brings the "unity" of the universe into play, proposing that behind all empirical appearances "there is a background continuous with the background of the brain." By unity, he doesn't mean a holistic universe, in which all things are interconnected; he simply means that the universe is a coherent system whose elements are compatible with one another. He arrives at this concept by the experience of the unity of our mind: "If the unity of a man's consciousness is not an illusion, there must be some corresponding unity in the relations of the mind-stuff, which is behind [the measurements]." How exciting is that: From our inner sense of unity, we infer the unity of the universe. The argument is that, if the universe was nothing but a pile of broken pieces of debris that didn't fit together, the unity of our mind would be impossible. Likewise, if the cosmic background is continuous with the rest, we can conclude from our own mind that the background of reality has mindlike properties.

The other day one of our friends had a birthday party, and somehow the conversation turned to Eddington's mind-stuff. My friends were polite, but they weren't convinced. "Are you telling us," one of them said, lifting her glass of wine, "that we are drinking thoughts?"

"Yes and no," I replied. "When we drink a nice glass of wine, it isn't the same as thinking about it, but the thoughtlike principles inside the molecules in your glass are responsible for the taste that you like."

So what does it mean that atoms have a thoughtlike background? It means exactly the same that it means for you and me: You have a body, and you can touch it, but you also have a mind that isn't tangible. Of course atoms have a mass, and it is the combination of the masses that you taste when you have a glass of wine. But what Eddington is saying is that atoms also have an inside that is thoughtlike, and the wine wouldn't taste the same without it. Because of the mindlike background of things we begin to think that the universe, too, has an inside and outside. The point is that matter isn't the only thing. There is something else: In previous parts of this book we called it the *virtual order*. The consistency or homogeneity of reality is that it has the same structure throughout: mindlike inside and material outside. Where there is matter, there is form.

Eddington warns us not to get carried away when we accept that an atom "has a nature capable of manifesting itself as mental activity." It is easy to attach details to this concept that don't apply. For example, Eddington says, we mustn't attribute to the background of atoms "the more specialized attributes of consciousness." But, he continued, that isn't needed to make it compatible with the nature of our mind.

In this way, step by step and carefully, Eddington is led to the conclusion that "the stuff of the world is mind-stuff." He hastens to add that the terms *mind* and *stuff* aren't perfect, but they are the best that he can use to describe the situation. "The mind-stuff of the world is, of course, something more general than our individual conscious minds; but we may think of its nature as not altogether foreign to the feelings in our consciousness."

Above we suggested that matter, energy, and form are equivalent. In this equivalence, the forms are primary in the sense that the ETs leave the realm of matter and enter the realm of forms whenever they are on their own. Moreover, the material structures of the

empirical world seem to be secondary also because they are actualizations of forms. In this sense, what really counts is the mind-stuff
of the universe. Mind matters! Eddington was fully aware that his
approach is alien to physics. "It is difficult for the matter-of-fact
physicist to accept the view that the substratum of everything is of
mental character." But we don't have a choice because there are no
"two avenues of approach to an understanding" of the nature of
reality. "We have only one approach, namely, through our direct
knowledge of mind.

"Consciousness is not sharply defined, but fades into subconsciousness; and beyond that we must postulate something indefi-
nite but yet continuous with our mental nature. This I take to be
the world-stuff."

The strength of this view is its consistency with the quantum
phenomena. We compare the mind-stuff "to our conscious feelings," Eddington concludes, "because, now that we are convinced
of the formal and symbolic character of the entities of physics, there
is nothing else to liken it to."

How the Cosmic Consciousness Reveals Itself in the Wholeness of the Universe

Eddington's mind-stuff meets all the conditions that an element
of the quantum world has to obey. It is mindlike, not matterlike.
It is beyond experience. It has the nature of a potentiality. Instead
of using the term *stuff,* we could describe it in terms of elements
of mind, or elementary thoughts, or ETs. The realm where these
thoughts exist is a mindlike realm of consciousness.

On issues of this kind, it is hard to find a consensus. As far as
we can see, the universe is a collection of material things: planets,
stars, galaxies, and so on. It doesn't seem to have a mind, where

we could locate its consciousness. In this situation, Menas Kafatos and Robert Nadeau have added a powerful argument, which they describe in their book *The Conscious Universe.*

For Kafatos and Nadeau the starting point is the aspect of the wholeness of the universe, as it appears in the quantum phenomena (see chapter 3). The wholeness "'implies' without being able to 'prove' that human consciousness participates in the life of the cosmos in ways that classical physics completely disallowed."

The state of wholeness in the universe doesn't appear to our senses because it is an aspect of the nonempirical realm of reality. All things and people aren't connected in their material structures, but in their transcendent roots. We have a good understanding of a state of wholeness, Kafatos and Nadeau point out, because it is a condition of our mind. If you consider your mind, I am sure you will agree that it gives you a feeling of wholeness. Its connectedness with other minds extends the wholeness to a transpersonal domain.

Kafatos and Nadeau's main argument for the importance of consciousness at the cosmic level amounts to this: If the universe is an indivisible wholeness, everything comes out of the wholeness and everything belongs to it, including our consciousness. Thus, consciousness must be a cosmic property, and it is possible to think that the universe is conscious.

The argument from wholeness is as simple as it is powerful, and you don't have to stop here. If everything that comes out of the wholeness belongs to the wholeness, "everything" means all things and phenomena, including the phenomenon of life. Like consciousness, life has come out of the wholeness and belongs to it: The universe is alive!

Some years ago, a good friend and coworker of mine, a prominent physicist and a passionate atheist, loved to tease me about the title of my first book, *In Search of Divine Reality.* "Ah!" he liked to say without ever taking a look at the book. "That's all wrong!" One day, in one of our discussions, I mentioned to him the argument from wholeness, and I saw that it stunned him. For a while he was quiet and thought about it, and then he said, "Let's talk about something else," and he never again wanted to talk about the search of divine reality.

At this point you can also see the fundamental differences between contemporary physics and yesterday's biology. In Darwinism, consciousness comes out of nothing. In quantum physics, nothing comes out of nothing, but everything comes out of the cosmic potentiality. That the consciousness in us came out of the wholeness means that its existence in us is the result of a logical process and not of a lottery.

Arguments for cosmic wholeness and consciousness won't try to force you to accept them, but they offer a choice: You can take it or leave it. One way to approach this situation is to ask what the alternatives are. The question isn't who is right and who is wrong, but whose story is the more likely one.

Having a choice is part of your potential. Interestingly, your choice in this case will define your potential: If you reduce the universe to a mechanical order, you also reduce to zero your potential of being creative. In the quantum universe, your potential is infinite.

How the Mindlike Background of the Universe Reveals Itself in the Principles of Our Mind

It is a characteristic aspect of our mind that it is somehow incomplete or open-ended. By this I mean that, in order to live in this world, we have to rely on operational principles or concepts, which we trust without being able to prove them. Somehow they exist in our mind and we use them as a basis of our life, but we have no idea where they are coming from. What I like to suggest is that these operational principles appear in our mind out of the mindlike background of the universe. They are actualizations of forms that exist in the realm of the cosmic potentiality, and they can appear in us because the universe is a mindlike wholeness and our mind is connected with it. Let me describe to you what this process is about.

Take any simple but general statement about the world and try to prove it. For example, you may remember how your physics teacher told you that at normal atmospheric pressure, water boils at 212 degrees Fahrenheit. That sounds like a reasonable proposition, but how would you go about proving it?

All right, you can take a pot of water, put it on the stove, submerge a thermometer, and watch it. Sure enough, as soon as the temperature reaches 212°F, the water boils. But what does that prove? It proves that this particular pot of water boiled at 212°F. But the statement that you wanted to prove was of a general kind that claimed that water *always* boils at 212°F when the atmospheric pressure is normal. Now how would you go about proving that? The answer is you can't!

You can take a million pots of water and determine their boiling points. Sure enough, all of them boil at 212°F. But that doesn't prove that the next one, pot one million and one, won't boil at 205°F. The problem is that you can't prove any general statement. That includes all laws of physics, because they are fashioned as general laws. You can test such laws, as science philosopher Karl Popper pointed out, but you can't prove them. Since all scientific statements are of this kind, the eighteenth-century British philosopher David Hume accused science of being a mere faith. Thus, in order to survive in this world, we have to rely on principles that we can't prove. It is like our mind is incomplete. It is advanced enough to use the principles that we need to survive in this world, but it doesn't have the tools to prove them. It's like the mind of a child.

If this is the first time that you've heard about this, your reaction might be: "I am doing quite well, thank you! Why worry about principles of science and whether they can be proved or not? They seem to work for me. So what is the problem?"

Well, actually, there is no problem here, but an opportunity to learn something about yourself, your mind, and your connection with the universe.

One of the aspects of the world that we have to learn as we grow up is the uninterrupted existence of things, called *object permanence* by the Swiss psychologist Jean Piaget. Piaget discovered that a newborn baby needs nearly one year to learn that things continue to exist when they are out of sight. This is the excitement of the peekaboo game that babies love to play. When you hide behind the sofa and then appear again, your baby can't control its delight over this miracle. When a magician lifts a hat that he put over a rabbit and the rabbit is gone, that's nothing special. For a baby, that's normal: When things are out of sight, they cease to exist. The magic would be that, when the magician lifts the hat, the rabbit is there *again.*

Piaget also discovered that children who don't learn object permanence will develop emotional disorders. Since object permanence is so significant for us, we should know why we believe in the uninterrupted existence of things. Normally we believe in principles because we can prove them. So how would you go about proving object permanence? Why don't you follow the eighteenth-century Irish philosopher George Berkeley, who claimed that "to be is to be perceived," meaning that things, when they aren't in someone's mind, will vanish from the universe?

I am sure that you have sometimes seen a cat in a playful mood, hopping around a room, up on a table and down on a chair, and under the sofa. After some time she comes back out again, and looks at you, happy. Have you ever wondered why you think that the two cats are the same: the one that vanished under the sofa and the other one that, at some later time, came out again? What if I told you that cats behind sofas vanish from the universe?

Actually, I am not suggesting that! But the point is that you don't have any visible evidence that your cat, when it goes out of sight, doesn't vanish from the visible world, like ETs do: You just take it for granted. And in the same way there are lots of other principles—in fact, all the general principles of science—that you take for granted but treat like proven facts.

The *principle of causality* is another such principle that we need in life but can't prove. Your car doesn't all of a sudden leave your garage and drive to the grocery store. Your dirty laundry doesn't all of a sudden drop itself in the washing machine and take a shower. Things like this don't happen because our world is a causal world: Except for quantum jumps, nothing happens without a cause.

We need the principle of causality because without it we couldn't make any sense of the world. For example, you trust what you can see of the world because you believe that the signals that arrive at your senses have a cause. But, David Hume claimed, we don't have a single experience of a causal event. Causality isn't a principle of nature but a habit of the human mind.

Imagine that you are playing tennis. You serve the ball to your opponent, who hits it back to you. When you are asked to describe this process, you will probably say that, by hitting the ball with his racket, your opponent caused it to return to you. But that isn't what you see. You see that the flying ball touches a racket and, at that exact same moment, changes its direction of flight. You *observe* a temporal conjunction, but you *infer* a causal connection. "No object ever discovers," Hume wrote, "the causes which produced it, or the effects which will arise from it."

So you see where all this is taking us. There are principles in your mind that you need to live in this world, but they aren't your own because you can't prove them. They form a system program of your mind, like the operational system of your computer. The program seems to be reliable, but you have no idea why that is so because you don't know where its principles are coming from. Since your mind didn't write this program, we have to suspect that the human mind is logically incomplete or open-ended. This brings up a question: What is our mind open to? My answer is that it is open to the realm of forms in the cosmic potentiality. That realm can bring forms into our mind because its nature is mindlike and we are connected with it.

Conclusion: The mysterious principles in our mind that help us to get around in this world are yet another sign that the universe is mindlike and a wholeness, because it can connect with our mind.

How the Mindlike Background of the Universe Reveals Itself in Synchronistic Events

Synchronicity is a concept introduced by Carl Jung to describe the simultaneous appearance of two or more events that are connected in meaning but not in their visible causes. Jung's German term *sinngemäße Koinzidenz* is difficult to translate into a short term. It means something like a "coincidence according to meaning." For simplicity, the usual translation is "meaningful coincidence of two or more events," and their synchronicity means that, as Jung defines it, "something other than the probability of chance is involved" in the simultaneous appearance.

Such a coincidence is found, for example, when you dream of an unexpected event that a little later actually happens. As Jung describes it, we are dealing with synchronicity whenever a mental state in an observer coincides with an external event sharing the same meaning. When the president of the United States appears in your dream and tells you that he wants to meet you and the next day you happen to see him in a restaurant—that is a case of synchronicity. A dream and an external event don't have a common visible cause. Synchronicity is at its strangest, Jung points out, when the inner mental state coincides with an external event that "takes place outside the observer's field of perception, i.e., at a distance, and only verifiable afterward."

Synchronistic events occur so frequently that they can't be explained away as merely random events. Too many people have thought of a name only to have that person telephone the next moment, or have seen a word in their mind that five seconds later was spoken by someone sitting next to them. Having foreknowledge of future events is also called precognition, clairvoyance, or telepathy. In such events our mind seems to be able to transcend space and time.

Within the framework of classical physics, no explanation can be given for how such phenomena can occur or what the means of communication are by which events distant in time and space suddenly become joined in our minds. The only thing that's clear is that the connection doesn't involve physical energy or forces, nor does it involve causality. "No one has yet succeeded," Jung writes, "in constructing a causal bridge between the elements making up a synchronistic coincidence." Nevertheless, Jung had no doubt that synchronicity was "based on some kind of principle, or on some property of the empirical world."

You probably know already what I am going to suggest at this point: If you direct the search for the basis of synchronistic events away from the "empirical world" and place it into the transempirical realm of forms, the realm of potentiality, the mindlike background, and the wholeness of the universe, you have a natural stage where synchronistic events can play. The connecting ground of synchronistic events is the nonempirical realm of forms in the cosmic potentiality. Turning the argument around: We should expect that synchronistic events will occur if the universe is a mindlike wholeness.

The forms in the cosmic realm of potentiality have two ways in which they can appear in the empirical world. They can appear as images in our mind or as structures in the material world. When a given form actualizes in both modes, then a subjective state in someone's mind is synchronous with an external event. Both the mental event and the empirical occurrence express the same meaning, because they actualize the same pattern of the realm of potentiality. That meaning is what Jung called "the common bridge" between them.

Thus, when the nature of quantum reality is taken into account, the pieces of the puzzle start falling into place. Instead of searching for a common cause in synchronous events, we only need to concede that they are based on processes in the realm of potentiality, where causality doesn't apply. The connection exists in the nonempirical realm of the world.

Jung lived at a time when the quantum phenomena weren't known to the extent that they are known today, but his intuitions about synchronicity are up to date. He was right, for example, that synchronicity points to the need of modifying our everyday view of the world that classical physics insisted to be true. Synchronicity is a powerful argument for the existence of a hidden part of reality whose nature is an indivisible and mindlike wholeness. As Jung described it, either "the psyche cannot be localized in space, or space is relative to the psyche."

Synchronicity can appear in more than a few events and it can involve more than a small number of people. Our history has known bursts of synchronistic developments that involved countless people, entire countries, and even the entire world. Take a look, for example, at the following sequence of events:

In 1900, Max Planck founded quantum physics, and Sigmund Freud invented psychoanalysis. In 1903, Henry Ford founded his motor company, and the Wright Brothers succeeded in the first human motor flight. In 1905, Albert Einstein developed the theory of relativity, and the first exhibit of modern art opened in Paris, with Fauvist paintings by André Derain and Henri Matisse. In 1907, Georges Braque and Pablo Picasso created the first Cubist paintings. In 1910, Arnold Schönberg composed the first piece of atonal music. In 1912, Wassily Kandinsky invented abstract painting. In 1913, Franz Kafka published his short stories. In 1914, James Joyce wrote *The Dubliners* and the First World War began; 1917 was the year of the Russian Revolution.

Each of these dates marks the appearance of a dislocation in a cultural structure. All aspects of life were involved. When the scientists changed their views of the world, the artists did the same, and the social order broke down. What is so striking about these events is that they were synchronistic: They had a common meaning, leading the world into the same direction.

Before 1900, for example, people were preoccupied with the plane of the human senses and the conscious experience of the world; the unconscious and the nonempirical were irrelevant. Paintings were realistic and used the techniques of perspective to simulate space as we see it. In the same spirit, classical physics was focused on the visible surface of things, using visible phenomena to describe the world, such as waves, particles, and fields. Everybody can take a look and understand these phenomena. Classical physics, writes physicist Roland Omnès, is "explanatory physics, where reality is visually represented in a way that can be fully grasped by intuition." In addition, the thinking of the age prior to 1900 was focused on absolute principles that couldn't be doubted, such as causality, determinism, and the objectivity of reality. All of these were thought to be unquestionably and fundamentally rooted in nature, because nothing else made "sense."

After 1900, in a burst of synchronistic events, there was a loss of realism in the arts, and physicists discovered the nonempirical realm of reality. At the same time, psychologists discovered the significance of the unconscious for our life, and artists discovered that the essence of things isn't found on their visible surface, but behind it: They didn't want to paint things the way they looked, but the way they were. Paintings became "evocative" and stopped being "reproductive," as Werner Haftmann describes this trend in his exciting book *Painting in the 20th Century.*

The eternal pointlike particle was the symbol on which classical physics explained the world. As it turned out, there is no such thing. In the European Renaissance, paintings were perspectivistic. The structure of such paintings was also based on a pointlike unit called the *infinite point* of perspective. When physicists gave up the pointlike particle, artists abandoned the infinite point of perspective in nonperspective, abstract paintings. In the same way, musical compositions lost their "key note" in Arnold Schönberg's atonal music. If you want to know what happened to the worldview of physics in the twentieth century, all you have to do is to take a look at modern art and listen to Schönberg's music.

When the classical perception of reality collapsed, it was a sign that the political order was soon to follow. The First World War spelled the end of several empires that had existed for hundreds of years, and the social order was turned upside down in the Communist Revolution of 1917 and the German Revolution of 1918. There was an earthquake in the "ground of existence" as Haftmann called it.

Even more subtle parallels can be found. The decades before 1900 were an era of spectacular archaeological findings. Between 1870 and 1883, for example, Heinrich Schliemann excavated Troy. Philosopher Jean Gebser compared the excavation of Troy to the "excavation of the unconscious" by Freud.

There is true synchronicity in all these events. The minds of different people were connected by meaning but not by visible ties. There was very little direct contact between the scientists and the artists involved in the process. The physicists weren't led to their discoveries by looking at the paintings of modern artists. Except for some isolated exceptions, the artists weren't inspired by the new physics to paint the way they began to paint. And the politicians and the military didn't have modern art and quantum physics in mind when they began the incredible slaughter of the First World War. Rather, all these minds were connected in the wholeness of the mindlike realm of the cosmic potentiality: *The cosmic spirit was at work*.

By affecting mental processes, the realm of potentiality in the universe has shown that it has mindlike properties. By affecting the entire world, it has shown that it is everywhere. The mental isn't localized in the universe in isolated islands, but its thoughts form an invisible ocean spread out through the entire world.

PERENNIAL PHILOSOPHY AS A SPECIAL FORM OF SYNCHRONICITY

When you take a look at how human thinking has evolved in our history, one aspect will immediately strike you: There are recurring ideas about the order of the world and human nature that are so deeply rooted in us that they constantly emerge in the thinking of people in different ages, different cultures, and different parts of the world. Already thousands of years ago, the Indian sages knew about this phenomenon; they called it *Sanatana Dharma*. The term can be translated in many ways: In a fascinating book, *In Search of the Cradle of Civilization,* Georg Feuerstein, Subhash Kak, and David Frawley translate *Sanatana Dharma* as "Perennial Wisdom," "Eternal Teaching," and "Global Spiritual Tradition." In the history of Western philosophy, Sanatana Dharma was introduced as *perennial philosophy*.

I consider perennial philosophy as a special form of synchronicity: It is the independent appearance, without any visible connection, of identical thoughts in different minds, at different times, and in different parts of the world. Darwinists will probably say that the recurring concepts are somehow genetically programmed. Instead, I take the phenomenon as a sign that human minds have been connected with a cosmic field at all times and everywhere in the world. The connection is possible because consciousness is a cosmic property and our individual minds are connected with it. In our discussions of the quantum phenomena we have found numerous concepts that can be understood as concepts of perennial philosophy. The Pythagoreans, for example, were the first to propose that all things are numbers. Aristotle was the first Western philosopher to introduce the concept of potentiality as a modality of being. The concept of the atom was invented by Leucippus and Democritus around 400 BCE, a long time before John Dalton introduced it to chemistry in the eighteenth century CE. Similarly, countless philosophers, including the Indian sages, Parmenides, Plotinus, and Hegel, thought that "all is one" a long time before the concept of wholeness appeared in quantum physics. Many other examples can be found, such as Plato's thesis that atoms are mathematical forms and Kant's view that there is a part of reality—the world of the noumena—that we can't see. In some cases, including Anaximander's apeiron and the Shaivist concept of spanda, the congruence of the spiritual and the physical is so complex and detailed that it is striking.

In their attempt to understand the nature of things, the early Greek philosophers started out by searching for a primeval substance or element out of which all things are made. Some thought that this element was earth; others thought that it was water, or air and fire. Searching for a primeval substance as the source of all being was the beginning of the materialistic European tradition of thinking.

In this situation and in contrast to this trend, Anaximander of Miletus, a sixth-century-BCE Greek philosopher, developed the concept of the *apeiron*. The origin of the world, he said, and the cause of all being isn't any material stuff that we know; it isn't anything like earth or water; it is something indeterminate and infinite and, as the historian of philosophy Hans Joachim Störig describes

Anaximander's definition of the apeiron, "cold and warm, dry and wet have separated out of it."

Hirschberger has discussed how the concept of the apeiron implies the existence of a nonempirical realm of reality that underlies the visible world. Hirschberger's argument is that something that is indeterminate and infinite is logically undefined and must be infinite both in space and time; that is, it is eternal and omnipresent. Something that is totally indeterminate can't be anything empirical. It can't be anything material because if it has empirical qualities it isn't completely indeterminate. If it consists of matter, it must have some empirical properties. So, the apeiron is necessarily nonempirical, nonmaterial, and, since everything that exists comes out of it, it has the nature of a potentiality: Anaximander's apeiron is the archaic model of Heisenberg's quantum potentiality. As Störig describes it: "Constantly new worlds are emerging out of the Indeterminate-Infinite and return into it."

Like Heisenberg's quantum potentiality, Schrödinger's wave mechanics, too, has ancient roots.

When Erwin Schrödinger postulated that the electrons in atoms are waves, the scientific community was stunned, because Schrödinger's theory implies that the visible world is created out of a realm of invisible waves. Few scientists were aware of the fact that Schrödinger's wave mechanics revived an ancient spiritual principle: the concept of waves as a principle of cosmic creativity.

The up-and-down and back-and-forth of waves create a vibrational motion, like the swinging of a pendulum. When we think of waves in a symbolic way, they are often associated with a boundless ocean, like the ocean of the divine, of consciousness, or of infinite creativity. In Kashmir Shaivism, a Hindu philosophy of the eighth century CE, cosmic vibrations were thought to be a principle of cosmic creativity.

In Kashmir Shaivism, *spanda* denotes subtle vibrations, waves, or throbs of the universe. These waves aren't visible in a physical medium, like water waves or airwaves, but are "vibrations in the divine." In this spiritual tradition, origin and basis of the manifested world were seen in a divine consciousness that has the nature of vibrations. The empirical world is an emanation out of a realm of waves, or an ocean of potentiality.

You can find an exciting description of these concepts in Paul Eduardo Muller-Ortega's book *The Triadic Heart of Śiva*: "That self-consciousness in the Heart in which the entire universe without remainder is dissolved," Muller-Ortega writes, "is called in the authoritative texts the vibration (spanda), and more precisely, the ultimate vibration (samanya-spanda), and its nature is an overflowing of the Self."

It is fascinating how the concept of a cosmic consciousness appeared in connection with spanda as well as the quantum waves. As Muller-Ortega explains, spanda "is the wave of the ocean of consciousness, without which there is no consciousness at all. . . . This consciousness is the essence of all." Without this cosmic consciousness our own consciousness wouldn't exist.

The world of Kashmir Shaivism is nondualistic. All things and phenomena are united in the cosmic consciousness or its waves, in which the manifested world is created. "The Ultimate," Muller-Ortega explains, "is *spanda*: it vibrates, it expands and contracts; it manifests and reabsorbs." And in quantum physics, material particles constantly come out of the realm of potentiality waves and dissolve again in it: They expand and contract, they manifest and reabsorb.

Quantum physics doesn't understand the realm of potentiality as an "ocean of consciousness," and every reference to the numinous is alien to it. Nevertheless, the forms in the realm of potentiality have mindlike properties, like spanda. These similarities weren't intended by the physicists, but they show, nevertheless, to what extent physics has moved science into the context of spirituality. Spanda is, as Muller-Ortega writes, "internal dynamism." It is the source of the movement that "results in the process of manifestation." It is "full of bliss and yet suffering occurs; it plays a game of hide-and-seek with itself in which ignorance alternates with knowledge, and in which enjoyment and liberation can coincide."

Apart from the fascinating connotations with contemporary physics, I find the poetry of this understanding of the world very moving. "The silence of the Supreme is shot through with a creative tension," writes Muller-Ortega, "a primordial urge, an impelling force. This force is the *shakti,* the power of the Ultimate." It is the power that sets up the "vibration that characterizes consciousness, and which allows consciousness to be the foundation and essence of all manifest reality."

So what are the lessons to learn from perennial philosophy? Do we have to believe now that Schrödinger's waves are throbs in the divine?

I don't think that the lessons are so specific, but they are of a general kind. We realize, for example, that science isn't so serene and disconnected from our spiritual thinking—conscious or unconscious—as the scientists always claimed. It doesn't seem very likely to me, either, that the concepts of perennial philosophy are somehow wired into our genes. I don't see, for example, how the awareness of the apeiron enhances my biological fitness.

I take the longevity of ideas as a sign of their verity. Many theories are constantly invented, but only those that correctly describe a true aspect of the world survive in the memory of humanity. The same is found in the arts. Lots of things are constantly being painted and sung about. But if it doesn't catch a true aspect of the world, no painting, poem, or song will endure.

The most important lesson from perennial philosophy is that consciousness means cosmic connectedness. Your mind is connected with a transpersonal domain. It follows that your potential isn't restricted to the activities of your genes—it is much more. It is the creative tension, the primordial urge, the impelling force: it is the spanda in you and, therefore, it is infinite. Just try it out!

THE HOLISTIC ASPECTS OF CONSCIOUSNESS

If consciousness is a cosmic property, it makes sense that the universe is a wholeness, because consciousness is a holistic phenomenon.

It is worthwhile to take a closer look at this word, *consciousness.* Everybody uses this word, but we rarely spend any time on finding out where it comes from and what exactly it is telling us.

The roots of the word are in the Latin "conscientia," which is derived from the Latin words *cum,* meaning "with"; and *scientia,* meaning "knowledge" or "science." So consciousness is a state of knowing or science together with something or someone. But what does that mean? With what or whom are we sharing what we are conscious of? My consciousness feels like a very private affair. Whenever you speak of your consciousness, you refer to it as yours. That doesn't sound like something that you are sharing with anybody else.

Words mean more than they say. We have seen it again and again: Words can brainwash. There is no such thing as objectivity in thinking and discussing. You can be objective only to the extent that your language allows you to. Words activate inner images that affect you in ways you aren't even aware of. So when the word that you use to refer to your consciousness implies that you aren't alone in this process, it is meaningful to find out who that external agent might be.

Through thousands of years, at different times, in different parts of the world and in different minds and cultures, the idea has emerged again and again that a cosmic spirit exists and is thinking in us. Originally, this idea had religious roots, and that may tempt you to shrug it off as the usual propaganda of the usual suspects. However, the case isn't that simple, because the same idea is now emerging in physics.

In previous chapters we looked at many examples in which suggestions of spirit appear in the quantum world. If a cosmic mind is involved in the thoughtlike background of every atom in the world, as Eddington thought, it is impossible to think that it has nothing to do with the molecular structures of our brain, and therefore with our consciousness. Thus, it makes sense that consciousness is thinking together with the mindlike background.

Atoms weren't discovered in physics or chemistry; originally they were mythological, even spiritual, units. As Carl Jung explains it, the concept of the atom didn't arise on the basis of any empirical evidence or observations of atomic phenomena, "but on a 'mythological' conception of smallest particles, which, as the smallest animated parts, the soul-atoms, are known even to the still Paleolithic inhabitants of central Australia." Does it matter for us today where the concept of atoms came from? You might say that you couldn't care less, because you don't accept atoms for mythological reasons but as scientific facts. However, what Jung is telling us—and I believe that he is right—is that the source of the images in us that help us to understand the world is the same now as it was in ancient times.

Evolutionary biologists like to point out that our mind evolved as an accidental function of our neural system, and, once it was there, we kept it because it enhanced our survival skills. There is no doubt that a well-functioning mind will enhance your biological fitness, but that may be a secondary effect. Instead, the primary function of your mind is its ability to serve as an outlet for the cosmic spirit. Your mind is a tool by which the cosmic consciousness can actualize its potentiality in the empirical world.

WHAT DOES THE COSMIC MIND-STUFF GET US?

Coming out of physics, the concept of spirit or consciousness as a cosmic principle has been a surprise, and you may wonder what you should make of it. One way people often deal with a surprise is to shrug it off as irrelevant. Thus, it is meaningful to ask what the cosmic mind-stuff, if it really exists, will get us.

In matters of this kind there are no proofs, and you have a choice: Is it more likely that the arrival of intelligence in the world is the result of a lottery, or is it the actualization of a cosmic potential? You know my answer; you have to find your own. Whatever it will be, you have to realize that the consequences are enormous.

Eddington argued that if the structure of the universe is homogeneous, a mindlike background should exist outside of our mind. You can turn this argument around and think that if the background of the universe isn't mindlike, the feeling that you have of your own mind is an illusion. So these are the options. To give your mind the status of something beyond illusion, you need a cosmic background. In this way the question of whether or not the cosmic mind-stuff exists is the same as the question of human dignity and a meaningful life. A homogeneous universe offers a chance of dignity. If your mind is an illusion, everything connected with it is an illusion: your values, your principles, your love, your joy of life, your care for others and the excitement of your inner potential. In short: the cosmic mind-stuff defines your humanity.

So far we have described four unexpected aspects of the world that the quantum phenomena have revealed: the nonmaterial basis of matter in probability waves; the nonempirical basis of the empirical world in virtual states; the wholeness of the universe; and the universe's mindlike properties. With this, we have to proceed into new territory, shifting our focus away from the universe and onto ourselves: What exactly do the quantum phenomena mean to our life? Specifically, is there a chance that they can help us to better our life?

The human significance of the quantum phenomena is three-fold: It leads us to a new understanding of our history; it has important advice for how we should live in this world; and it has some suggestions for an impending mutation of our consciousness.

PART TWO

LIVING
in
THIS
REALITY

Chapter 5

DARWIN WAS WRONG:
EVOLUTION NEEDS QUANTUM SELECTION
AND COOPERATION

"It must be recognized that monist-materialism leads to a rejection or devaluation of all that matters in life."

—JOHN C. ECCLES

*I*f there is one conclusion that we can draw from the previous chapters, then it is this: Reality is difficult to understand because it hides behind its visible surface. You have to be smart not to fall for its tricks. In some sense the classical sciences—Newton's physics and Darwin's biology—and the view of the world that emerged from them weren't smart, no matter how technically brilliant and successful they were, because they always told us that the visible surface is the thing to watch if we want to know what the world is about. Now, quantum science reveals a new level of reality and offers a new beginning.

Watching the visible surface of things is what biologists have been doing for centuries, and we have learned a lot in this way. Unfortunately, their successes have led the biologists to think that the visible surface is all that matters and that, therefore, the quantum properties of the molecules at the basis of life have no significance for biology. Specifically, the neo-Darwinists—that is, the representatives of an updated version of Darwin's theory—typically claim that quantum science has no significance for the evolution of life.

This assertion is false. It matters that molecular processes at the basis of evolution are ruled by quantum laws. The mutation of a gene, for example, is a quantum process, and the virtual states of individual molecules are involved. This is the meaning of the short term: Biological selection is quantum selection. In addition, Darwinism, or its modern version, neo-Darwinism, paints a picture of life that is in complete contrast with the nature of reality as it appears to us in the quantum phenomena.

In this chapter we will focus on this conflict between quantum science and the Darwinian view of the world because, beyond scientific dispute, this conflict is important for each one of us. Neo-Darwinism isn't only a scientific theory but also a way of life. Because its view of the world is the wrong view, it has led humanity to a wrong way of life and the globe into a crisis. To find a way out of this crisis, understanding where neo-Darwinism goes wrong in the quantum world is of significance for all of us.

LIVING OUTSIDE THE ORDER OF THE UNIVERSE

We live in this universe and our life evolved in it. Is its order our order, or do we have an order of our own? Is it more likely that we are different from the rest of the world, or does our nature reflect the nature of the universe?

For some reason contemporary biologists insist that we exist outside of the universal order, in "total solitude," in "fundamental isolation," and "on the boundary of an alien world," as Jacques Monod described it. Monod was a prominent French biologist and Nobel laureate. We have to take it seriously when he says that man lives "like a gypsy" in an alien world. The same frame of mind is at work when biologists claim that the principles of quantum physics are of no significance for biology.

Darwin developed his theory of evolution in the age of classical

physics and in the spirit of his time—that is, the spirit of materialism and mechanism, which claims that all things can be explained
by the mechanical properties of solid material particles. By adopt-
ing Darwin's theses, modern biology placed itself within that same
paradigm and outside of the spirit of quantum physics and chemistry, even though biology is based on their laws.

In his popular book *Finding Darwin's God,* biologist Kenneth
Miller describes in detail how science deals "with the material"
and how science is "a form of practical, applied materialism"; he
also describes the view that "all that Darwin did was to show that
mechanism and materialism applied to biology, too."

Kenneth Miller is a prominent biologist, and we should respect-
fully consider his views. But there is a problem in that quantum
theory doesn't allow a mechanistic and materialistic description of
the world. So someone has to find a way to reconcile the mechanism and materialism of biology with the quantum nature of its
molecular basis, which is of particular significance for the evolution
of life.

Biologists often argue that the molecules of biological systems
are too large to be quantum systems. Thus, they claim, quantum
theory doesn't apply to biology. However, when you cut your finger,
what do you see? You see the red color of your blood, which is the
color of the hemoglobin in it. Hemoglobin is a giant molecule. It
is more than thirty thousand times heavier than a hydrogen molecule. Why does it look red? It looks red because it can absorb only
certain wavelengths out of a mixture of waves in sunlight; or, as we
described in chapter 2, it absorbs only certain *energy quanta.* The
absorption of quanta of light by a molecule can only happen in
one way: The molecule must make a transition from one quantum
state to another. Within the framework of contemporary physics,
light absorption can't be explained in any other way. It follows that,
even though hemoglobin is a giant molecule, it exists in quantum
states: It belongs to the quantum world and isn't excused from it.

We can find many examples of this kind that show that even the largest molecules of biology are quantum systems. If you are like many people, you are probably carefully watching your diet. The food that you eat has to be balanced. Among the various food groups, proteins contain some of the largest molecules that occur in nature.

In the 1980s the first so-called ab initio quantum chemical calculations of the structures of peptide molecules were performed in my research group. *Ab initio* means "from first principles." In computational chemistry, ab initio methods calculate the properties of molecules from the basic properties of atoms and their electrons.

Peptides are the building blocks of proteins. As it turned out, our calculations of peptides could be used to predict details of the structures of proteins that were so precise that they had never been observed before. The protein chemists to whom we talked about this typically waved their hands, pointing out that using quantum chemical methods to study the properties of proteins was a joke. But why had these details never been observed? Because, at that time, the experimental techniques of the protein crystallographers, whose work it is to determine protein structures, weren't precise enough. However, during the decade following the first publication of our results, the experimental techniques developed considerably, more precise protein structures became available, and our calculations were found to be in perfect agreement with the new experiments.

There are many ways to calculate molecular structures. Some methods are based on classical physics in that they treat the atoms in molecules as solid classical particles. When such methods are applied to the structures of peptides, the details of the quantum structures aren't obtained. This shows that the structures of proteins are expressions of quantum effects. Proteins aren't, as biologist Michael Denton and his coworkers point out, "contingent assemblages of matter like Lego constructs, watches or other sorts of artifacts—where the parts are the primary things and pre-exist the whole"; rather, proteins are made up of structural units that act like the cog wheels of a clock that coil up to little balls outside of the mechanism. As many of our calculations showed, this is so because the coiled-up shapes of isolated peptides are much more stable than the shapes that peptides have to assume in proteins. Because of this it is practically impossible, as Michael Denton and his friends explain, to find a convincing mechanism for the stepwise evolution of intermediate protein forms. Proteins are true quantum systems, and we should treat them with respect. Segregating things is a trademark of European classical science. Taking things apart and putting the parts into separate boxes is a neat and comfortable way to deal with nature. "Science works because it is based on causality," writes Kenneth Miller. "Once you understand a process, even a complex one, you can reduce it to the mechanistic sum of its parts." We can take this statement as a passionate appeal for logical clarity and objectivity. Unfortunately, at the molecular basis of biology there is no "mechanistic sum," and causality doesn't always apply, because single quantum events have no apparent causes. If we accept Miller's definition, this means that dealing with quantum systems can't be the business of science.

Segregation is the passion of the mechanistic mind. Eventually, that passion led to a dogmatic system affecting all aspects of life, including moral, public, and economic order. In the same way in which the molecules of biology are taken outside of the realm of other molecules, there is a disconnectedness of the arts from the natural sciences; of philosophy from the practical life. There is a constant conflict between science and religion; between our rational nature and spiritual nature. All of these phenomena are expressions of the mechanistic mind-set: They are in conflict with the wholeness of the world, and, for that matter, with a wholesome life.

How the Concept of Quantum Selection Arises in the Mutation of Genes

From Aristotle to Leibniz and Newton, a peculiar principle of European thinking has been that *nature doesn't make jumps*. For some reason Charles Darwin accepted this principle in his theory of evolution. The quantum phenomena are now showing that this principle is meaningless: Nature does make jumps. As a matter of fact, nature makes only jumps—namely, quantum jumps. The not-making-jumps principle is an example of how intuitive biases play a role in science. They can play a useful role, but you can never know when they will lead you to an error.

There is a whole system of such biases, and often one inspires another. Akin to the bias that nature doesn't make jumps is the bias that nature is causal. The principle of causality states that nothing happens without a material cause. Indeed, when you consider the things around you, nothing does happen without a cause, and any assumption of the opposite seems wrong. The furniture in your living room doesn't all of a sudden appear in your bathroom. However, when we step down to the level of molecules, all of a sudden the scenery changes: Molecules can and do act without being prompted by a mechanical cause.

In his book *The Blind Watchmaker,* evolutionary biologist Richard Dawkins discusses the significance of causation in genetic processes, specifically in the mutations of genes. It is often said that mutations happen by chance. The title of Jacques Monod's famous book is *Chance and Necessity,* meaning that mutations of genes happen by chance, while the process of copying genes is ruled by the necessity of invariance.

Dawkins discusses the role of chance in terms of the randomness of mutations. There is a way to say that mutations are random, but not without cause. This seems to be a contradiction, but you will see that it isn't when randomness is defined. "The question of whether mutation is really random is not a trivial question," Dawkins writes. And he explains that mutations are "random" only in the sense that there is "no general bias towards bodily improvement." At the same time, randomness mustn't be confused with having no cause, because even random mutations can be *caused* by something. In Dawkins's words: "Mutations are caused by definite physical events; they don't just spontaneously happen."

This is what I mean by the importance of biases in science. Biases such as "everything has a cause and nothing happens spontaneously" seem to have an empirical basis. But, as we have said before, reality is deceptive: Below its visible surface, it has some tricks up its sleeves. This notion, specifically, that mutations "don't just spontaneously happen" disregards an important degree of freedom that molecules have: They can undergo spontaneous changes of state that have no mechanical cause.

A mutation of a gene occurs in a chemical reaction of a specific gene molecule. All chemical reactions involve the virtual states of molecular systems. Quantum jumps into empty states are needed for a reaction to proceed. Like all state changes, the state transitions that occur during mutations can also be spontaneous. This means that they can occur without a visible cause. A simple example is found in the *depurination* of DNA molecules, which is one of the mechanisms that can lead to the mutation of a gene.

In living cells, gene molecules—such as DNA—constantly collide with water molecules. In depurination reactions a water molecule collides with DNA in such a way that it will push a group of atoms, called a *base,* out of the DNA, and then attach itself at the spot in the DNA that the leaving base left empty. You can compare this to a situation in which one of your friends kicks you out of your job in your company and then takes your place.

In a nutshell: In a depurination reaction a DNA molecule replaces one of its parts with a water molecule when the two interact in a collision. That process can start a sequence of events that will lead to a mutation.

This is how the notion of *states* comes into play in such reactions. Before anything happens, the initial state consists of the unchanged DNA and a water molecule that swims through the soup of the cell on its own. Call this "state A." At the end of the reaction, the final state, call it "state B," consists of the modified DNA plus the base that got kicked out and is now happily swimming through the soup on its own.

Here is the crucial point: To get from state A to state B, the reacting molecules must form what quantum chemists call a *transition state*. In the transition state all of the participants have loose ties with one another in a single state. The DNA molecule, its base, and the water molecule all form a loose alliance in a single system—you might say they are flirting with one another in a threesome—but they haven't made up their mind yet whether to proceed with the reaction or to drop back to the starting state. The important point is that all the state transitions involved in this process are quantum jumps. That is, they are *spontaneous* or *indeterminate,* and they have no cause. When the water molecule and the DNA first meet, they may jump into the transition state or they may not, and the water will stay in the soup. What will actually happen is entirely indeterminate. The jump from transition state into product state, too, is completely indeterminate: The threesome may decide to jump into state B or they may not, instead jumping back to state A. It is a general property of quantum jumps that they can occur without any causes.

In all quantum processes single, individual events are unpredictable. It would be a serious departure from the quantum norm if nature were to excuse genetic processes from this rule. Indeterminacy is what quantum processes are about, and physical reality doesn't make exceptions for biological molecules. Even in cases in which we say that a mutation is caused by something, such as a virus or X-rays, the underlying quantum nature of the process isn't suspended. For example, when a DNA molecule collides with an X-ray photon, it may absorb its energy, and then it may not. In each single case there is a probability for the absorption to occur and a probability that it doesn't occur. Even when a probability is overwhelmingly high or overwhelmingly low, there is an essential difference in quality between events that are causal and events that are ruled by chance.

You can also think of a mutation in terms of a mountain range and its valleys. In depurination reactions, for example, there is a valley for the DNA and water before they react, and another valley for the DNA with the water attached to it. As in a real mountain range, to get from one valley to another, you have to climb a hill. That same rule applies to reacting molecules: They have to climb a hill of potential energy to get from state A to state B. The top of the hill represents the transition state. After arriving at the top, the DNA-water-leaving-base threesome may decide to drop downhill into the valley of state B, and then the reaction will be final. Or the threesome may decide that they don't like state B after all and drop back into state A. The important point is that the course of events is unpredictable because of quantum indeterminacy: When the reaction actually occurs, it hasn't been caused by anything.

To be cautious, we say that spontaneous mutations have no causes in the *empirical* world. This means they have no visible, material, energetic, or mechanical causes. Whether some influences outside of our space-time exist and twiddle with a gene molecule, we can't know. I doubt if a materialist, mechanist, and atheist scientist would be happy to consider this option.

Some physicists have seriously considered such possibilities. In the 1930s, for example, Sir James Hopwood Jeans, a prominent astronomer, mathematician, and physicist, proposed that "the minutest phenomena of nature do not admit of representation in the space-time framework at all." Only some of the phenomena of nature can be described within the space and time of our universe; others can't. Among the latter, Jeans continued, "we have, for instance, already tentatively pictured consciousness as something outside the continuum."

It isn't the topic of this chapter, but it is amazing how often aspects of consciousness as a cosmic property emerge in these debates.

Spontaneous mutations aren't caused in any way, but in some sense they are *biased*. They are biased because transitions from a given state are usually possible to many other states for which the transition probabilities are different. Differences in transition probabilities will favor the selection of some states over others. This is like the transition probabilities observed by spectroscopists. It is a form of selection that we can call *quantum selection*. Nobody has any information on whether quantum selection can bias the process of evolution, such as favoring the direction of higher complexity. But the fact that a sort of quantum selection exists at the molecular level shows that natural selection isn't alone. Somehow the evolutionary progression is affected by the properties of the quantum states involved in mutations.

Paul Adrien Maurice Dirac, one of the great pioneers of quantum physics, has proposed that, in quantum jumps, a "choice" is made "on the part of nature." Physicist Henry Stapp suggests that "a 'choice' is defined by any fixing of an aspect of the universe that isn't fixed by the known laws of nature." You can think that the realm of unknown laws is the same as the nonempirical realm of reality that we have talked about above.

This is what is meant by "quantum selection." In spontaneous mutations the properties of virtual states come into play. The center of genetic changes is shifted out of the realm of material things—the genes—into the transempirical realm of potentiality, where new possibilities arise in nonempirical principles, which we can't know. Nobody can state a priori that such principles are irrelevant for biology. Genes—lumps of matter—aren't the masters of life. They aren't the authors of messages but just the actualizations of messages out of the mindlike, cosmic realm of potentiality: the realm of the cosmic spirit, which has found, in the genes, yet another way to express its virtual order in the empirical world.

Without the Quantum Properties of DNA There Would Be No Biological Evolution

Let's say that one day a friend is visiting you and asks you to do her a favor. She hands you a box with all kinds of springs, cogwheels, gears, screws, bolts, and a collection of clock casings and housings, in which the gears can be anchored. She also gives you a finished clock and asks you to make a copy for her.

Your friend is such a nice person. Copying a clock wasn't what you had in mind for that day, but it seems to be important to her, so you sit down and start copying the working clock that she gave you. How are you going to go about this?

To begin, you select out of the collection of clock housings the one that is identical with the housing of the clock that you are about to copy. Next you pick the right spring and the cogwheels and bolts, as you find them in the clock that you have to copy, and you start screwing them together, always using the assembled clock as a template.

The point of this story is that, in copying a clock, there is no room for variation or error. If you selected the wrong housing, the right cogwheels wouldn't fit. In the right housing, the wrong cogwheels wouldn't fit. If you picked a single wrong gear, you probably couldn't even screw it into its anchors. If you could, anyway, the next gears wouldn't fit.

That same situation applies to genes. If gene molecules were what Michael Denton and his coworkers call Lego-like structures made up of solid interlocking bricks, or if they were made up, like clocks, of interdependent cogwheels that must be screwed together with the help of a template, there would be little or no room for variation when copies were made, and mutations would be rare, if not impossible.

DNA is a large molecule of a type chemists call *polymers*. The name derives from the Greek and means something like "many parts." Polymer molecules are typically long chains of many smaller molecular units—their parts—that are tied together. The smaller molecules that form DNA are called *nucleotides*.

In the same way in which a certain number of cents make a dollar, a collection of nucleotides make DNA.

In the same way in which there are cents, nickels, dimes, and quarters, there are different types of nucleotides. Four or five of them are the most important ones in forming the chains of DNA that are currently in use in living cells. The sequence of the different types of nucleotides is what defines a specific DNA.

With this, we are all set to consider how the quantum properties of molecules play a role in the synthesis of genes. When a DNA molecule is synthesized, one nucleotide after another is attached to the growing chain in a chemical reaction. At each step of the synthesis, the sequence of events is similar to what we considered above for the mutation of a gene. That is, there is a starting state of reactants and a catalyst whose function it is to facilitate the chemical transition. As before, to get from state A, the state of the reactants, to state B, the state of the products, the system must pass through a transition state: the hill between two valleys. The reactants are the unfinished chain of nucleotides and the single nucleotide that has to be added. In the transition state the reactants form a loose alliance. The state of the products contains the original chain of nucleotides elongated by one.

The important point is this: At each step of the synthesis of DNA the quantum rules apply. A growing chain, the catalyst and a single nucleotide *may* form a transition state, or maybe not. From the transition state they *may* jump to the product state, or maybe not. Nothing determines what happens in each state. What happens on the road stays on the road! Apart from that, at each point of the chain elongation, when a growing chain is supposed to add a nucleotide of type 1, it may form instead a transition state in which it gets entangled with a nucleotide of type 2 or type 4. When that happens, the interaction may lead to an actual bond and a mutation has occurred. As before: In each mutation there is an element of caprice, and a choice is made on the part of nature.

This account is simplified, but it illustrates the *fundamental* importance of the quantum nature of genetic processes. The quantum nature *encourages* mutations because its rules are both lawful *and* playful, and allow for a certain creativity of the process. Copying a mechanical clock isn't a creative process.

This is part of the beauty of the quantum world: It is lawful and playful. In contrast, the classical world is lawful and predictable—that is, boring.

Neurobiologist Joachim Bauer has come to exactly the same conclusion. He notes that, on the one hand, biological processes are lawful, "on the other, all biological systems display considerable tolerance . . . so that processes in the single case can take a varied course."

NOT FROM NOTHING, NOT BY ERROR, NOT IN BLINDNESS: BUT BY THE COSMIC POTENTIALITY

In his paper "Three Levels of Emergent Phenomena," biologist Terrence Deacon writes about the evolution of life: "Creating something from nothing is an important part of what the universe is about, and some of the most intriguing examples of this curious process are what define us as living thinking beings."

Often when we think about the nature of this world, we come

up with ideas that are so fascinating that they take our mind in their grip. The idea that the complex forms of life come out of nothing is such an idea.

If you think about it, the Darwinian system of thought makes sense. Species obviously change. Their changes seem random, meaning that they aren't driven by the purposes of a designer. Moreover, in the variation of species, the visible order emerges out of nothing that we can see! Accordingly, it makes sense that biologists describe the process of copying genes as "blind," and mutations as "errors."

Here, again, when we step down to the molecular level, the anthropomorphic descriptions of things come to a sudden halt.

Nobody can deny that we are chemical reactors. At every moment your life is based on the chemical properties of the molecules in your cells. So if you claim that living chemical reactors will evolve new structures out of nothing, you must find the molecular basis for this process. That is, you must find a mechanism by which molecules create new forms out of nothing.

But there is no such mechanism, because molecules are quantum systems. As we saw in chapter 2, all that a molecule can do is jump from an occupied state to an empty state. It can't jump into nothing. The restructuring of a DNA molecule, for example, can be understood as the actualization of one of its virtual states. It is true that, when several atoms or molecules interact, new states will emerge in the interaction that don't exist in the isolated atoms or molecules. But these new states, too, don't come out of nothing; they emerge in a predictable way out of the interacting potentialities of the individual atoms or molecules. A hydrogen molecule, for example, isn't created out of nothing, but it is the actualization of a virtual state that belongs to the potentialities of interacting hydrogen atoms. The same principle applies to the emergence of all new structures in living organisms.

We can bring these considerations to a short form: Chemical

systems can form new structures not out of nothing but out of the interacting potentialities of their components.

From the times of the Celts to the Middle Ages, the inhabitants of northern Spain believed that the westernmost point of Europe, a cape in the Atlantic Ocean, was the end of the world. So they called it Cape of the End of the World, or *Cabo Fisterra* in the Galician language. It was general knowledge that, if you sailed west off that cape, after some time you would abruptly fall off the edge of the earth and into nothing.

Creating something out of nothing implies that such a cape exists in the world of molecules, as if when two molecules react, they proceed to State Fisterra, from which they drop into nothing, and in the process a new life-form emerges. But dropping off the deep end isn't a degree of freedom that molecules have. There is no Cabo Fisterra in chemical reactions.

How often, when something unexpected happened to you, did you say something like: "That came out of the blue! Like out of nothing!" By reacting in this way, you didn't really mean *blue* and *nothing*. You meant an unexpected event came out of something, but you didn't know what that was. In the same way we must understand that the complex order in the biosphere comes out of a realm of reality that we don't really know: that is, the cosmic potentiality.

This is such an important aspect of biological evolution that we should put it into the form of a general principle: The complex order of the biosphere appears in the actualization of a virtual order that already exists in the cosmic potentiality before it is an empirical order. *Virtual state actualization* (VSA) is the mechanism by which a new order evolves in the world. If you think about it, the principle applies to everything, not only life.

Related to the concept of the evolution from nothing is the view that life evolves in accumulations of "errors." In his paper "The

Unknown, the Unknowable, and Free Will as a Religious Obligation," biologist Robert Pollack writes, "Facts from science tell us . . .
that our species—with all our appreciation of ourselves as unique
individuals—is not the creation of design, but the result of accu-
mulated errors."

Here, again, the anthropomorphic view has great appeal; and,
again, the molecular level will spoil the fun. It is a delightful
thought that complex systems come into being because someone or
something doesn't watch out and makes an error. But in the emergence of complex order by VSA, the concept of error doesn't occur.
It only *seems* that errors are made, because the quantum processes
are indeterminate. When a specific stretch of DNA is copied and
the copy turns out different from the original, there are no flaws
nor faults; the system simply actualized a state that was available
to it. Quantum entities don't think about getting things right; they
just populate quantum states.

The appeal to *nothing* and *error* shows how the classical perspective can give the process of evolution a twist that it doesn't really
have. The same has to be said about Jacques Monod's description
of genetic processes in terms of *chance and necessity*. The concept
of necessity is completely off the mark: In the synthesis of single
gene molecules there is *no* necessity. All molecular syntheses are
quantum processes, and the outcome of the synthesis of a specific
molecule is inherently unpredictable. Many outcomes are usually
possible with different probabilities that depend on the waveforms
of the states involved. Thus, in a very basic sense, the units of natural selection aren't stretches of chromosomes but the waveforms of
quantum states that actualize in chromosomes. Life-forms are cor-
related with quantum waveforms.

So, not from nothing, not by error, and not in blindness: but by
VSA. When Monod uses the concept of blindness in connection
with evolution, again he misses the point: "A *totally* blind process,"
he writes, "can by definition lead to anything; it can even lead to

vision itself." No, it can't. No process, blind or seeing, can arbitrarily lead to *anything*. It can lead only to something that is already contained in the cosmic potentiality. If vision weren't part of it, we wouldn't see a thing. This may be the reason why evolution hasn't led to mythical monsters, such as griffons or basilisks or humans with wings: Cosmic virtual states don't exist for such beasts. Do you realize what a boost of your biological fitness it would be if you had wings and could cross a street when the traffic light was red because you could simply fly to the other side?

Evolving Complexity by
Cosmic Virtual State Actualization

For some reason one hesitates to accept that in molecular quantum jumps a transcendent, mindlike, and virtual order expresses itself in the material world. We have a habit of putting the concepts of mind, transcendence, and spirituality into boxes of their own, to keep them apart from the rest of the world.

In chapter 3, I described arguments for the view that the universe is a coherent wholeness, in which all things are interconnected. If we adopt this view, then it is possible to think that all empirical structures are somehow represented in the wholeness of the cosmic potentiality by a form, or a subspace of forms. In that case, the idea of DNA molecules may have existed in the cosmic potentiality a long time before actual DNA molecules appeared in the empirical world. Chances are that even you existed as a subspace of forms in the cosmic potentiality a long time before you were born, and the subspace that is you will continue to exist in the potentiality after you die. Of course, complex structures don't just jump out of the cosmic potentiality in one swoop; a process is needed—a process of growth and evolution. We could call CVSA—cosmic virtual state actualization—the process by which complex systems emerge out of the cosmic potentiality. For lack of a better term we have implied that "representation in the cosmic potentiality" is a *state*. But it is a mere manner of speaking: We can't really know the nature of the nonempirical world.

Throughout his career, French paleontologist and Jesuit priest Pierre Teilhard de Chardin emphasized the power of "the psychic" and of "thought in the stuff of the universe." The concept of CVSA revives such views, because virtual states are mindlike, not matterlike, and form the mindlike background of the universe. The transempirical order underlying all of reality is, at the same time, *immanent,* because it is contained in the things, and *transcendent,* because it isn't part of the visible world.

It is interesting to note that chance plays a role in both CVSA and Darwinism. But in Darwinism, the evolving order is *created* by chance; it is a "noise" that natural selection will turn into "music," as Monod described it. In contrast, in the quantum evolution of life, the music is part of an ongoing cosmic concert. Chance is at work in the quantum jumping, too: A jump may occur or not, and it may land on this state or that. But the order of the states on which the jumping will land has nothing to do with chance.

THE EVOLUTION OF LIFE IN A HOLISTIC UNIVERSE

Let's stop for a second and consider what kind of principles and behaviors you would expect for the inhabitants of a universe whose basis is an undivided wholeness. Would you not expect that, in such a universe, unifying principles such as love and bonding, cooperation and care, and mutual support should play an important role?

The empirical world has different qualities from the underlying realm of the cosmic potentiality. It is a dualistic world, irrevocably structured in polarities, including good and evil, love and hatred, war and peace. In the human world, specifically, such polarities are an essential part of our nature: Without the choice between opposing poles, freedom and humanness would have no meaning. The choice between good and evil is really a choice between a life that is in contact with the wholeness and instructed by it, and a life in which the connection with the One has been suspended. Only if we choose to act in the spirit of the One can we make contact with the One.

The point is that there are *polarities* in our world, not *singularities*. We have a choice! In his theory of evolution Charles Darwin took away our right to choose. He made the mistake of selecting the negative poles—war, aggression, and selfishness—as the *primary* principles of progress. In doing so, he drove us out of the cosmic context, like Adam and Eve were driven out of paradise. Darwin's species have no choice: Either they are on the top, or they are eaten by someone on the top.

Life is evolving within the order of reality. We must think that we will fare best when we are in contact with the One and live in accordance with its principles. But a life motivated by the Darwinian principles of segregation, aggression, and selfishness as its *primary* virtues isn't connected with the One. How could life evolve in processes of segregation, in which the disadvantaged are marginalized and eliminated, if the nature of the universe is that of a wholeness?

As it turns out, important stages in the history of evolution didn't depend on aggression, as various researchers have pointed out, but were made possible by biological cooperation. Genes aren't "selfish," as Richard Dawkins claims. The fundamental biological principles of the genome are, as Joachim Bauer explains, "cooperativeness, communication, and creativity."

The formation of the first primitive cells is an example in which the Darwinian view of aggression clashes with the emerging view of cooperation in biology. Let's compare Richard Dawkins's account with the descriptions given by biologist Rhawn Joseph, neurobiologist Joachim Bauer, and microbiologist Carl Woese.

Without any doubt, Richard Dawkins's description makes a wonderful story. He envisions that, in the "primeval soup," somehow the first molecules formed, which were able to replicate: He calls them "replicator molecules," or, simply, "replicators," who "were the ancestors of life."

As time went on, more and more replicators came into being, different varieties formed, and they began to compete for food. "There was a struggle for existence among replicator varieties," Dawkins reports. "These proto-carnivores simultaneously obtained food and removed competing rivals." Apart from being delightful, Dawkins's account is impeccably Darwinist: "Any mis-copying that resulted in a new higher level of stability, or a new way of reducing the stability of rivals, was automatically preserved and multiplied." Here you have it again: Making errors has its rewards; it leads to fitness.

In important schools of contemporary biology, the view of the beginnings of life is entirely different. An increasing number of scientists are exploring the possibility that the beginning of life wasn't in the ploys of "proto-carnivores," but, as Joseph points out, in the cooperation between different systems, which *shared* information in gene transfer processes. Innovation wasn't the achievement of a few ruthless operators that cannibalized the competition, but a communal project. As Carl Woese points out, the cell designs needed in support of the evolution of life were so complex that they couldn't be achieved by individuals, and not even by single species. "It is the community as a whole, the ecosystem, which evolves."

It seems that, as Joseph and Bauer explain, every advance of complexity in the evolution of life was made possible by the willingness to cooperate and to share. As Joachim Bauer describes it:

Development was typically not that of "lone warriors (neither lone-warrior individuals, nor lone-warrior species)," but it was the development "of biological systems." This is exactly what you might expect for a process that evolves in a holistic universe.

Of course, Darwinism also acknowledges the importance of cooperation and altruism. But, in Darwinism, altruism is not a *primary* principle of a holistic system; it evolves in secondary processes as "optimized strategies in the struggle for survival."

When you consider Richard Dawkins's definition of altruism, you can't help but think that his altruism isn't what we normally mean by this term. As he describes it: An animal or a gene "is said to be altruistic if it has the effect (not purpose) of promoting the welfare of another entity, at the expense of its own welfare."

Many other basic assumptions of Darwinism are contradicted by recent research. For example, genetic changes aren't completely random, but, as John Cairns, Julie Overbaugh, and Stephan Miller, Harvard School of Public Health, explain, "cells may have mechanisms for choosing which mutations will occur." Furthermore, as Joachim Bauer and American paleontologist and evolutionary biologist Stephen Jay Gould point out, in the history of evolution, innovation wasn't achieved in small steps but proceeded, as Bauer describes it, in "surges" in which "the self-modification of organisms follows identifiable principles, which are laid out in the biological system itself."

"Species formations are the work of an inherent dynamic," Bauer concludes, "which is laid out in a given genome. Living systems are not just victims of evolution, but they act in it."

Being "laid out in the biological system itself" is the same as being anchored in the potentiality of the system.

Seeds as Centers of Potentiality

In an exciting paper, Rhawn Joseph describes a scenario of evolution that is completely outside of the Darwinian dogma. In this description Joseph combines discoveries of contemporary biology with the thesis of Fred Hoyle and Chandra Wickramasinghe that life originally came to this planet from outer space, carried by cosmic debris, or dust and meteorites. The first "microbes to arrive on earth," Joseph describes, "contained the genetic machinery for growing the tree of life." They also changed the planet in a way that created the right living conditions, "altering the womb of the planet so this 'seed' could grow branches and bear fruit in the form of increasingly complex species."

The term *metamorphosis* derives from Greek and denotes a *transformation* of some sorts. (The Greek *morph* means "form," and the prefix *meta* has various meanings, among them "beyond.") In many cases a metamorphosis is a change in which hidden or seemingly numinous forces are at work. In biology the term is used to denote changes in life-forms that occur during the development of some animals, like when caterpillars turn into butterflies. When we look at such metamorphoses, we don't think that an evolutionary process is at work that evolves one type of animal from another; we explain them in terms of the activities of genes that can be turned on and off during the growth of an animal.

At the beginning of your life, one fertilized egg cell, the *zygote,* contained the potential for everything that could happen in your biological life. Out of this one single cell, all the other types of cells in your body were formed: the liver cells, kidney cells, muscle cells, and brain cells. We don't look at these cells like they are different animals or species that evolved by chance out of a community of cells and somehow survived in the struggle for life. In the same way, Rhawn Joseph proposes, we should look at the evolution of life.

Like the development of an embryo and like the metamorphoses of caterpillars, "'evolution' *is under genetic regulatory control.*" The first cells that existed on this planet, Joseph explains, "contained the genes and genetic information for altering the environment, the 'evolution' of multicellular eukaryotes, and the metamorphosis of all subsequent species and their extinction."

The early microbes that arrived on this planet from outer space were "genetic seeds." The tree of life grew out of them, like an apple tree can grow out of an apple seed. There is no process of "evolution"; there is only the actualization of the potentiality contained in the seeds before it is actualized.

The extinction of countless species is an unquestionable fact of the history of life on this planet. In Rhawn Joseph's analysis, the extinction of genes, cells, and entire species is also genetically programmed. When a tree grows and loses one of its branches, we don't look at this as the result of the aggression of the other branches that are striving for biological fitness. In the same way countless species fell off of the tree of life. "Like programmed cell death," Joseph explains, "extinction is often intrinsic to and necessary for the development, evolution, and metamorphosis of increasingly complex species."

Joseph describes an exciting aspect of the process when he says that life doesn't actualize outside of the cosmic order, but instead new species emerge in the interactions of existing species with the environment. Signals from the environment influence genetic activity. Genetic activity, in turn, can affect the environment, and environmental changes can influence what forms of life will actualize and what forms will be eliminated. For example, microbial activity at an earlier stage in the history of this planet changed the environment by increasing the concentration of oxygen in the global atmosphere.

Life is an expression of the cosmic potentiality. Thus I think that it is actualizing in many places in the universe. In this process life on Earth is, as Joseph describes it, "just a sample of life's evolutionary possibilities." The potentiality is that of a system. Ultimately the system includes the genes, the planet, the solar system, the galaxy: the cosmic potentiality. Joseph's account of the actualization of life doesn't rely on aggression as a primary virtue; it is in agreement with the principle of potentiality as a basis of reality; it doesn't have to assume that the laws of quantum chemistry don't apply to biology; it operates within the order of the universe and doesn't send us into exile to "the boundary of an alien world," as Monod thought: In short, Joseph's is an immensely reasonable account whose time has come.

The Interplay of Potentiality and Actuality in the Intelligence of Life

If the universe is an undivided wholeness, the potentiality that is at work in the evolution of life is the cosmic potentiality. In that case we must think that the cosmic potentiality hasn't only started that process but continues to be involved with it. This is possible because living systems are endowed with an intelligence that allows them to be active in both domains: potentiality and actuality.

All physical phenomena of the world are affected by forms in the realm of potentiality. Why should biological phenomena be exempt from this rule? Of course, accepting the involvement of a transempirical background is in conflict with the operational principles of traditional biology, because it implies that living cells aren't sensitive only to chemical and physical signals in their environment, but also to nonmaterial signals out of the realm of potentiality. In spite of this conflict, a growing number of scientists are beginning to consider this possibility. Informatics engineer Karl Goser, for example, views such processes as the defining difference between machines and biological systems: "Machines cannot receive information from a transcendent world."

VSA AND THE PRE-DARWINIAN CONCEPTION OF EVOLUTION BY NATURAL LAW

The concept of the evolution of life out of a realm of transcendent forms isn't new. In the nineteenth century, for example, the pre-Darwinian biologists believed that life is evolving out of a realm of Platonic forms; Richard Owen was one of the prominent biologists at the time who developed this view.

In chapter 2 we saw how the visible order of atoms and molecules appears out of a realm of virtual forms. From this simple fact we concluded that all of reality is like that: The empirical world appears out of a cosmic realm of potentiality—a realm of nonempirical and virtual but preestablished Platonic ideas.

Biologists prior to Darwin were inspired by the "unity which underlies the diversity of animal structures," as Owen described it, to propose that the structures of living organisms are determined by natural law in the same way in which other structures in the world, such as molecules and crystals, are determined by natural law. The pre-Darwinian biologists wanted to explain the diversity of life-forms in a lawful way, unifying physics and biology, and they searched for a special class of biological laws, the *Laws of Form*. You will find a detailed description of this concept in the paper by biochemist Michael J. Denton and his coworkers Craig J. Marshall and Michael Legge. In a general sense the concept means that only forms that are copies of forms that exist in the logical order of the world can appear in the biosphere. Thus, life-forms aren't created by chance or by the specific survival needs of a species.

The pre-Darwinian view of life was a Platonic view. Owen

described anatomy, for example, in terms of "archetypes," and the organization of living organisms was seen to rise "from the general to the particular." In contrast to Darwinism, Denton and his co-workers point out, the Platonic view considered form primary and function secondary. If life-forms are expressions of natural laws, one can think that the course of evolution has followed a lawful path as well. The pre-Darwinian biologists ultimately failed because they weren't able to identify the laws of form and how they operated in biology. Their ideas were before their time, because the concept of a nonempirical realm of physical reality wasn't a possibility of the classical sciences. Thus, the Platonic view was abruptly replaced by Darwin's theory. As Michael Denton and his colleagues describe it, "necessity was replaced by contingency and natural law was replaced by natural selection."

The VSA principle has revived the pre-Darwinian Platonic view of evolution. The actualization of empirical order out of the realm of potentiality is a law of forms. That is, only forms that are copies of preexistent forms in the transcendent part of the universe can appear in the empirical world.

WHY IT MATTERS

At this point you might say: "OK! Maybe Darwin was wrong, and maybe the rules of quantum physics apply to biology, or maybe they don't. So what? I am alive, my body is functioning, why do I care about what kind of physics applies in it?"

The answer is simple: You should care because Darwinism isn't just a scientific theory of biology. It is a way of life, and it is driving the world you live in into a crisis.

For example, when generations of children are raised on the doctrine that aggression and selfishness are the primary virtues of living systems, what kind of a world do you expect that will create? A world of peace populated with helpful, caring, and loving people?

Science isn't only about technology; it is also about a view of the world. Darwinism is materialism. The only thing that matters is matter. If Darwin has it right, you are a machine. Your emotions, hopes, and dreams are important only insofar as they serve the manipulations of your genes. Your genes use you, as Dawkins describes it, as a "vehicle" for their selfish purposes.

If you think that you are an expression of a cosmic order, and if you like to live a caring and loving life, Darwinism is telling you to think again. This is the attitude of the losers! The replicators in you, as Dawkins says, define the rationale of your existence: "They are in you and me; they created us, body and mind; and their preservation is the ultimate rationale for our existence." This isn't just science; this is a sort of religion that can be called the GG cult: the adoration of the "God Gene." "They have come a long way, those replicators," Dawkins explains. "Now they go by the name of genes, and we are their survival machines."

I hasten to point out that many essential aspects of Darwin's evolutionary theory are beyond any doubt, and they are discoveries of great significance. There is no doubt that, for example, the visible forms of life aren't constant but subject to change. There is no doubt that different life-forms share a common origin, and life didn't appear on this planet in accordance with the biblical account—that is, in seven days, starting on a Monday morning at 8 a.m. (assuming normal office hours). In addition to describing aspects of our evolution that we have to accept as true, Darwin's hypotheses had immensely beneficial effects. For example, they have done much to take humanity out of the grip of religious dogmatism, which has brought incredible suffering on the world.

At the same time, many details of Darwin's theory are under attack. We have already mentioned that there are reasons to believe that species formations didn't proceed in gradual steps, but in jumps. The timing of these jumps wasn't random but coincided with periods of global ecological stress. Furthermore, complex order doesn't emerge out of nothing and isn't created by chance; and the war of nature isn't the primary principle of progress, but at important steps cooperation was important. Darwin's biology is, like Newton's physics, a surface science. It describes the visible phenomena on the surface of reality quite well, but it isn't able to get to the roots of things. Apart from these technical problems, conceptual difficulties arise when Darwin's theory is used as a basis for a metaphysic of morals, as is done by sociobiologists.

Sociobiologists are scientists who believe that our social behavior, like that of all animals, is genetically coded. That is, they search in our biological roots for the principles of conduct that we accept in our life, meaning that our essential behaviors are evolutionary adaptations. If a behavior is an adaptation, it means that one of your ancestors was selected for it. It means that our values, including our moral values, are nothing but strategies that serve the selfish schemes of our genes.

In "The Approach of Sociobiology: The Evolution of Ethics," Michael Ruse and E. O. Wilson write that morality "is merely an adaptation put in place to further our reproductive ends." An adaptation in this context means a change in behavior for which our ancestors were selected in the process of natural selection because it enhanced their biological fitness. So if this is what morality is about, Ruse and Wilson explain, then the conclusion must be that "the basis of ethics does not lie in God's will—or in the metaphorical roots of evolution or any other part of the framework of the Universe. In an important sense, ethics as we understand it is an illusion fobbed off on us by our genes to get us to cooperate. It is without external grounding."

Many people are surprised by such thoughts. But once the Darwinian, or neo-Darwinist, view of life is accepted, the consequences described by Ruse and Wilson follow quite naturally; they are logically cogent, and we have to be grateful to Ruse and Wilson for pointing them out. The general conclusion is that our actions aren't guided by some higher code of ethics, as many people might think, but by our biology. As Ruse and Wilson explain it: "The way our biology enforces its ends is by making us think that there is an objective higher code, to which we are all subject."

If human values are merely strategies in a game aimed at getting the better of someone else's genes, then I can propose a shortcut that has some practical value because of its simplicity. In case of doubt, don't ask yourself "Is what I am doing good?" Rather, ask yourself "Is what I am doing good for my genes?"

In his later work, Michael Ruse further developed the views of sociobiology. "Our morality is put in place," he writes, "by our biology, to make us good social animals." And furthermore: "I do not accept objective moral properties." This leads him to the conclusion that morality is "a collective illusion of humankind, put in place by our genes in order to make us good cooperators."

At first encounter such views are so unexpected that they are in some sense delightful. They are like a fresh breeze in an often stuffy environment. But then you realize that the medication has its side effects. You realize that Darwin's hypothesis has an impact not only on biological research but on our life. And this is the reason it matters for all of us that the basis of Darwinist or neo-Darwinist thinking is scientifically not sound: *The wrong view of reality leads to the wrong life.*

In European history, Darwinist thinking has had a disastrous effect. Neo-Darwinists today love to discuss controversial issues with Christian fundamentalists. It is like setting up a smoke screen, because the sorry theses of fundamentalists have been taken care of a long time ago. Why should they still be discussed? Instead of chasing red herrings, we should discuss whether there are any lessons to be learned from the fact that Darwinist writers in Germany were able to pave the way for the racist policies and the horrible crimes of Nazi Germany. Richard Weikart describes in his book *From Darwin to Hitler* how German Darwinist writers and intellectuals at the end of the nineteenth and the beginning of the twentieth century developed the concept of "the good of the race" as a standard for public order. In doing so, Weikart describes, their writings became a source of inspiration for the "Evolutionary Ethics, Eugenics, and Racism" of the Nazis.

In the theses of the sociobiologists, the catastrophic impact that Darwin's hypothesis has had on the fate of humanity comes to the fore, even though I am sure Darwin didn't intend this. If there are no "objective moral principles," if the world is nothing but an arena for the schemes of selfish genes in the struggle for the annihilation of species, then the characteristic crimes of our time—the countless wars, the destruction of the environment, the adoration of violence and brutality, the manipulation of the public, the abuse of women, and the exploitation of labor—all are justified in one fell swoop as legitimate means in the struggle for survival. In such a world the current economic system, in which the greed of the fittest can ruin the rest of humanity, also follows quite naturally.

In the same way in which Jesus can't be blamed for the inquisition and its autos-da-fé, it isn't Darwin's fault that his disciples have erected a totalitarian system on the theses of their master. But it will be our fault if we wait any longer to reclaim the understanding of human beings and their values and put them again on a realistic basis.

You have to admire the logical consistency of the world of classical science. Everything fit together: physics, biology, ethics, metaphysics, and public order. Darwin's biology was created in the spirit of its time. It was a late justification of colonialism and the crimes of early capitalism. If you can sail out in the world and conquer its countries, you should do it. If you can enslave or manipulate the people of your country to make some money, you should do it: It is all a part of the Darwinian game.

The examples show the task ahead of us: We, too, have to develop a logically consistent system of physics, biology, metaphysics, ethics, and more: psychology, medicine, political science, international relations, and the views of our spirituality. All of these have to be joined in an integrated system. But to be wholesome, our system has to be in agreement with the nature of the world. And that means with the quantum world.

EVOLUTION BY ADAPTATION TO THE FORMS OF THE COSMIC POTENTIALITY

Take a second to consider this question: If, at one time in the history of life, consciousness stepped out of the cosmic wholeness and became a part of us, would it not make sense to think that we were still connected with it? Or would it make more sense if the connection were made just once, and then the receiver was hung up? However you look at the matter, it seems reasonable to think that the human mind isn't self-contained or self-sustained, but connected with a mindlike wholeness. "We can 'infer'" Menas Kafatos and Robert Nadeau suggest, "that human consciousness 'partakes' or 'participates in' the conscious universe."

As I have made sure to emphasize, science can't *prove* that the universe is conscious. At the same time, the numerous suggestions by serious scientists, including Bohm, Dürr, Eddington, Fischbeck, Jeans, Kafatos, Lipton, Nadeau, and me, that a cosmic spirit exists can't all be shrugged off as signs of dementia in these authors. It makes more sense to conclude, as psychiatrist Brian Lancaster has done, that "consciousness amounts to a fundamental property, irreducible to other features of the universe such as energy or matter."

Regardless of what we can prove, it is worthwhile to consider what the activities of a cosmic realm of consciousness could mean for us if it did, indeed, exist in the universe. What can we think its role could be in the evolution of life? My suggestion is that, if consciousness belongs to the cosmic potentiality, it will look for living organisms as a means to manifest itself in the empirical world. For all we know, it is possible to think that the forms in the cosmic potentiality may be evolving together with us to ever increasing complexity.

In Darwin's evolutionary theory, the primary process is the adaptation of individuals and species to their environment. That process is driven by purely material needs. But if the basis of reality is mindlike, we have to expect that the needs of life aren't purely material but also mental. Considering the power of our mind, this leads me to think that the primary principle of evolution is the adaptation of individuals and species to forms of the cosmic potentiality. This may sound like a rather abstract principle, but it is really quite simple.

Adaptation to forms is a process by which living organisms evolve the ability to receive, understand, and actualize increasingly complex signals out of the cosmic potentiality. The reception of signals needs a receiver that has been adapted in the right way. The forms that your mind can actualize are more complex than the forms that your pet understands. And the forms to which an animal is adapted are more complex than the forms that a molecule understands. This is what I mean by adaptation to forms of the cosmic potentiality.

You can also look at this process from the point of view of the cosmic consciousness. The adaptation of minds to forms is a process in which the cosmic consciousness creates increasingly complex bases or ports through which it can enter the empirical world. We can't see any reason or purpose in this process. But we do see that it has a direction: It is directed toward increasing complexity. For all we know, the interaction may even be mutual, and the cosmic consciousness may in some sense profit from complex thoughts that we develop.

In the adaptation to cosmic forms, species aren't selected for their aggression in the struggle of life, but for their thoughtfulness. This process is automatically an adaptation to the environment, as well, because both the order of the external world and the principles of our mind are actualizations of the same cosmic order.

You might argue that we don't really know how the cosmic potentiality interacts with the human brain. But that problem applies to all of our interactions with the nonempirical world. We don't know, for example, how your mind can affect the material brain, but there is no doubt that it is doing it. The evolution of consciousness is just another process of the spontaneous self-actualization of virtual order, comparable to its actualization at the atomic level, and there is no mechanistic description of it. It occurs quite naturally, spontaneously, and without any stress. It also occurs with a certain creativity and playfulness. There is much beauty in this world—more than is needed for survival, as though it was the work of a playful artist.

"Animals develop queer adaptations," Deepak Chopra remarked in one of our recent discussions, "like the giant anteater's long tongue, when it is obvious that all around this creature, other animals without that adaptation survive perfectly well." The cosmic spirit is playful. It acts more like an artist than an engineer. "What benefits certain terns to fly thousands of miles to nest when other birds nest close to home?" Chopra asked. "Why should the echidna and platypus continue to lay eggs, millions of years after other mammals gave that up? There are lots of colorful examples of this type."

Many people are amazed when they learn that intelligence is a necessary property at all levels of life. Even single-celled organisms display the ability of a planning control and intelligence that is comparable to the intentionality of our mind. "Each cell is an intelligent being," writes cell biologist Bruce Lipton. "Like humans, single cells analyze thousands of stimuli from the microenvironment they inhabit." In this context the facts of psychosomatic medicine, which show that the mind can affect the biochemistry of the body, are also relevant. If consciousness is a cosmic property, it is reasonable to think that the intelligence of biological systems is a form of its manifestation.

Spirit is everywhere. We have a consciousness because the universe has a consciousness. It wouldn't make sense to believe that the evolution of life isn't affected by that consciousness.

How the New View of the World Can Help

Take a second and consider how incompatible these two views are: a) The nature of reality is that of an undivided wholeness; and b) The principles of living systems are those of selfishness and separateness outside of the wholeness. The conflict is obvious and intolerable. We can't go on living with this kind of schizophrenia!

In this situation the discovery, in physics, of a transcendent cosmic order is of the utmost significance: It offers a way out of "a robber's life" as Plato called it. In his book *For a Civil Society,* Hans-Peter Dürr describes how the awareness of quantum reality can help us build a kinder world and a society whose order is based on community, not adversity; on cooperation, not competition. "We are not 'stuck' with an innate viciously competitive nature," writes Bruce Lipton in his book *The Biology of Belief.* Instead, "survival of the most loving is the only ethic that will ensure not only a healthy personal life but also a healthy planet."

Lipton's views illustrate the importance of the potential in you. You aren't a cogwheel in a cosmic clock, but more like a thought in the cosmic potentiality. The potential in you allows you to choose what kind of world you will create and live in. In the cosmic potentiality there are no values, just forms. The values emerge in the process of the actualization of the forms. Human kindness or madness: You decide. The key is in you!

At this point it is worthwhile to consider again the synchronicity of the various developments. At a time when physicists are discovering the wholeness of reality, biologists are discovering the importance of connecting principles for the evolution of life. In addition, recent investigations in neurobiology have shown that, as Joachim Bauer describes, human beings aren't wired for aggression and strife, but for social bonding and successful relationships.

When you look inside yourself, it is easy to decide. Do you enjoy constant stress and conflict, or do you cherish love and friendship? Our neurobiological motivation system rewards care and love, while aggression leads to somatic stress. "The core of all motivation is the desire to find and to give inter-human recognition, respect, attention or affection. From the point of view of neurobiology, we are beings which are designed for social resonance and cooperation," writes Joachim Bauer. Since our brain turns "psychology into biology," Bauer concludes, people who live in the stress of constant readiness to bite and to attack have a different biochemistry than people who meet the world with care and humaneness.

It is like we said: The wrong view of the world will soon lead to the wrong life—and stress-related illnesses.

As a metaphysic of human values, the Darwinian worldview and that of its modern syntheses have become unrealistic, because they deny the interconnectedness in the wholeness and the mindlike nature of the transcendent cosmic background. The time has come to return to a realistic view of humanity and to arrange our public order in accordance with the nature of reality. For example, we might try a society that isn't based on competition but cooperation, and international relations that aren't based on military aggression but mutual support. Whatever we try, it should be different from the unenlightened life that the sociobiologists are offering.

At this point Pierre Teilhard de Chardin's view comes to mind that life is under pressure everywhere "to burst from the smallest crack no matter where in the universe." He further thought that, "once it appeared, it is incapable of not using every opportunity and means to arrive at the extreme of everything that it can accomplish, externally of Complexity, and internally of Consciousness."

I can't help but point out this striking congruence: Teilhard speaks of the appearance of life like of a process of actualization. You can take his description to mean that, once it has found a way into the empirical world, life is under pressure to actualize its potential: You are this life and its pressure is your pressure, and the potential that Teilhard is speaking of is your potential. The pressure is the urge that you feel in you to find and to actualize this mysterious creativity that has to do with your happiness. If you neglect your cosmic task, there won't be punishment in hell, just the frustration of a boring and wasted life.

Basically, biological evolution is the selection of forms out of the cosmic potentiality. Quantum selection, unlike Darwinian selection, isn't committed to selfishness but to principles of wholeness. Human conduct can bring values into actuality that aren't based on the Darwinian virtues of selfishness, deception, and ruthlessness. What I propose is that the principles that appear in our thinking are reflections of the universal order. If this is so, the adaptation involved in moral behavior that the sociobiologists refer to is the capacity of the mind to comprehend the significance of universal principles. In the same way in which we evolved the capacity to understand the universal principles of physics we evolved the capacity to understand the universal principles in ethics.

Thus, all of a sudden, the discussion of issues in science has unavoidably led us into the realm of human values.

Since we live in this universe and our life evolved in it, we belong to it, and its order is our order. We have a body and mind, because the universe has a body—the material world—and a mind—the realm of potentiality. Having body and mind is part of the rules of the game.

From a journey through the landscape of the quantum world in the first part of this book, we found our way in this chapter to specifically human matters: important aspects of our biology and, with them, questions of morality. Can the nature of reality really have some advice for us as to how we should live together? In the next chapter we'll explore the possible connections.

Chapter 6

World Ethos: Living in Harmony with the Order of the Universe

"Our natures are parts of the World-Whole. For that reason, the final goal is to live in accordance with Nature; that is, the life in accordance with our own nature as well as the nature of the universe. In such a life one undertakes nothing that the World-Reason (. . . really: the general law) forbids. World-Reason is the true reason (orthos logos), *which permeates everything and is one in essence with Zeus, who provides order to the universe and guides it."*

—Zeno of Citium, fourth century BCE,
cit. Hauskeller

Taking a walk through the quantum world we arrived, in the last chapter, at questions of our own origin and, with this, at questions of our social behavior. Is there a way we are supposed to act and, if so, why should we act that way? Or are there no rules to obey, and the fittest get their way?

Traditionally, people have answered such questions by referring to some higher authority, such as the will of God. But in a globalizing world that procedure doesn't work anymore. Why should a Muslim, for example, obey the same divine laws as a Christian, and why should an atheist want to obey any divine laws at all? If, in this situation, the order of the world could be a model for human order, offering guidance for how we should live, that would be very helpful.

The quantum world has, indeed, some advice for how to live together on this planet. In a short formula: In a holistic world in which all things and living beings are connected, we should do nothing to impair the other. In a wholeness it isn't smart to harm, hurt, or cheat, because if you cheat others, you ultimately cheat yourself. My discussion of these issues in this chapter will include thoughts from a recent essay that two friends of mine, Diogo Valadas Ponte, a Portuguese psychiatrist, and Sisir Roy, a theoretical physicist at the Indian Statistical Institute in Calcutta, published together with me in *Zygon: Journal of Religion and Science.* Apart from our thoughts, our cooperation is in itself an interesting symbol of an integrating world. At what time in our history would it have been possible for a European medic, an Indian physicist, and a physical chemist living in the United States to join their professional skills in discussing issues that were traditionally the privilege of philosophers, if not of clerics? This exchange of ideas significantly expanded earlier attempts to connect moral order with cosmic order.

Moral laws can and should be derived from cosmic order. Our understanding of the world should guide our life: Cosmic order is a model of human order. No God and no atheist principle can be violated by the suggestion that our moral rules have cosmic roots and that we should live in harmony with the order of reality. A life in agreement with cosmic order is an authentic life. The inauthentic life isn't worth living.

THE ACCIDENTAL HUMAN IN A DISCONNECTED WORLD

The nature of the European sciences is analytical: Things are understood by taking them apart. The technical success of this method has led to an obsessive passion for taking everything apart, applying the principle of segregation to everything.

In this way the arts and religion were separated from the sciences. International order was based on the separation of nations in never-ending conflicts and wars. In Newton's physics, each particle is centered onto itself in its eternal and immutable being. In Descartes's philosophy, mind was separated from matter. In Darwin's biology, each species varies its type by chance, separated from any lawful processes of nature. Jacques Monod's conclusion is unavoidable: We are strangers in an alien world that couldn't care less about our needs and hopes and crimes.

In this environment, David Hume was a master of segregation. We have already seen how he separated the order of the world from the principles that we use to describe the world. By this I mean that the principle of causality, for example, isn't a principle of nature but, as he said, a habit of the human mind. In the same way, Hume claimed, the laws of ethics are separated from the order of the world. Not only do the two not have anything to do with each other, Hume said, but something is wrong with you if you think otherwise.

This is Hume's famous *is-ought fallacy.* It is an error, he said, to try to derive principles of ethics, which tell us *what ought and ought not to be done,* from principles of nature, which tell us *what is and is not the case.* What the world is like has nothing to do with how you should act. There is a clean cut between rules that define cosmic order by describing what reality is, and rules that define human order by prescribing what we should do.

Hume's is-ought fallacy was hailed as a great discovery, even by idealist philosophers. But while this principle tells you not to let your view of the world affect your view of morality, it was itself inspired by Hume's view of the world. His worldview was that of the science of his time, which claimed that the world consists of isolated, lifeless, and mindless pieces of stuff that can do nothing but move about the universe in meaningless tracks. In this world there are mechanical principles at work and nothing else. Thus, indeed, any attempt to make a connection between human, moral order and cosmic, physical order is a "naturalistic fallacy," as G. E. Moore described it.

In chapter 5 we saw how contemporary biologists are completely immersed in the classical worldview. This should mean that they, too, would avoid any connection between moral order and natural order, in order to be logically consistent. Nevertheless, the sociobiologists fall victim to the naturalistic fallacy and allow their biology to inspire their moral theory. Without Darwin's biology there would be no sociobiology. We begin to suspect that worldview unavoidably affects the understanding of human values, no matter what the worldview is.

Searching for Human Values in the Depths of the Universe

Finding a generally acceptable basis for moral principles is a difficult task. In order to avoid these difficulties, I propose that we shouldn't ask ourselves what we *ought and ought not* to do, but what is *reasonable and meaningful* to do. Taking the question of our morality out of the realm of religion and into the realm of practicalities avoids Hume's is-ought fallacy and the problem of whose God we are supposed to obey. When we do that, we will no longer ask what we *ought or ought not to do* but whether what we are doing is *true or false, meaningful or meaningless, authentic or inauthentic,* and *wholesome or making us sick.* What I like to propose is that a moral life is a matter of prudence.

Take as a simple example the question whether it can be *meaningful* to live a life that is in conflict with the nature of reality. If you have a view of the world that isn't in agreement with the nature of the world, then that view is wrong. If you follow principles of life that are in conflict with the nature of the world, then these principles are wrong, too. Why would anybody want to live such a life? Similarly, can it be *good for your health* to live in a state of constant stress, always ready to attack and bite the next person in a selfish life that isn't focused on the wholeness of reality?

World ethos is the system of principles of conduct that proposes that only a life that is in harmony with the order of reality is meaningful or true; in view of the holistic nature of the world, such a life is a wholesome life. If the background of the universe is mindlike, we can trust that it can actualize moral principles inside us in the same way in which it actualizes material structures outside.

Thousands of years ago living in harmony with the order of the world wasn't considered a natural fallacy but an accepted principle of Greek philosophy. Around 300 BCE Zeno of Citium arrived in Athens from Cyprus and founded the school of the Stoics, which was to become a school of great influence in the history of Western philosophy. In Zeno's philosophy, a fundamental principle of ethics is "to live in accordance with Nature," as Michael Hauskeller describes it in his inspiring book on the history of ethics. And with this, Zeno didn't mean only our own nature, but also the nature of the universe. The ancient Greeks had a moral tradition that connected the concept of virtue with the development of the defining property of a thing. Within this tradition, as Hauskeller explains, virtue meant to achieve the "value-best state" of a thing. There is a Greek term for this kind of general excellence: *aretē*.

In the context of our current understanding of virtue, the concept of *aretē* at first sight seems to be missing the point. For example, you can apply the Greek principle to something such as a knife, and then you can say that the virtue of a knife is to be sharp. Similarly, you can apply it to an athlete, and then the virtue is to be strong. So this understanding of *aretē* seems to be more like the definition of a skill or a utility rather than a moral property. But when Zeno applied this principle to our human nature, the rules that followed were moral rules and a matter of human reason.

To Zeno, acting within the Greek tradition, the question of human virtue came down to determining what the defining property of human beings is. Our most characteristic property, Zeno said, is our reason. If there is anything that defines who you are, it is your ability to think and to think reasonably. Therefore, to develop our reason to its fullest power is our moral duty and the basis of a moral life. It is needless to say that, in order to develop our reason, we must live in harmony with the principles of reason.

At this point Zeno made an amazing turn to the cosmic. Even though it is our essence, he said, reason isn't the achievement of human beings but a gift on loan from the universe. Reason is to Zeno, Hauskeller explains, a "world-principle which in human beings rises to the level of reality—when it does rise to reality." The gift brings with it an obligation, a task that has to be fulfilled. Reason is like a seed that has been planted in us. You have to see to it that this seed will grow. Since your reason is an essential part of your potential, it is your obligation and the "divine will" in you to strive for the *aretē* or best possible actualization of your reason. The divine will in you is, at the same time, your own will. It is that same cosmic principle in you that we have encountered in this book in connection with the quantum phenomena: Your thinking is the thinking of the cosmic spirit in you.

It was clear to Zeno that this structure of the world defines a commitment that we have. As Michael Hauskeller describes it, it was Zeno's view that the first duty of every human being is to live in accordance with the nature of the universe. If you live in accordance with reason, you live in "harmony with nature."

"Our natures are parts of the World-Whole," Zeno writes. "Therefore, the final goal is a life in accordance with Nature; that is, a life in accordance with our own nature as well as the nature of the universe." Such a life leads to commitment and duty—Hauskeller cites Zeno's reasoning—because in such a life "one undertakes nothing, which the World-Reason (. . . really, the general law) would forbid."

Step by step, Zeno's arguments can be combined with our own: The virtuous life is in harmony with reason. But reason must be understood as a cosmic principle, of which the human reason is a part. The existence of a cosmic reason must mean that the universe is a process that has a meaning. We don't know what that meaning is, because it is hidden in the transempirical realm of the world. But you must think that you are a part of this process and that turning against the cosmic task is the same as turning against your own nature. Thus, living in harmony with nature doesn't mean only your personal nature, but also the nature of the universe, or, as Zeno writes, the "nature of the all-pervading World-Reason, which is law to all things." Living a life guided by reason isn't only in accordance with human nature, but, as Hauskeller concludes, in harmony with the nature of the cosmos.

Every time I think of Zeno, I have the feeling of a mystery. Thousands of years ago this Greek philosopher came up with a view of ethics that appeared in my own mind—a long time before I had ever heard of Zeno—in connection with the phenomena of quantum physics, when I was writing *In Search of Divine Reality*. These views are not common or ordinary; many people may find them strange. Physicist Stanley Klein, for example, comments: "Schäfer's discussion of cosmic morality and hope goes overboard for my taste."

Socrates believed that to live in accordance with the essence of things is the basis of a moral life. Since it is hard to live in accordance with the essence of something that you don't understand, developing an understanding of the world is one of your most important tasks. This is exactly what the quantum phenomena help you to do.

A WHOLESOME LIFE IN THE WHOLENESS
OF THE UNIVERSE

Compare these two worlds: Darwin's world of aggression, of the war of nature, of predators and the life of the fittest; and the quantum world, a world of wholeness, bonding, and mutual support. Now consider this question: How can it be possible that our view of the world has nothing to do with our way of life? I propose that the structure of reality as it reveals itself in the quantum phenomena offers valuable advice for how we should live. *The good life is in harmony with the order of reality.*

The first aspect of reality that is significant for our way of life is the fact that physical reality doesn't have only a material realm, but also a nonmaterial background. In a metaphorical way we have said that the universe has a body (the empirical world of material things) and a mind (the realm of nonmaterial forms). That structure of the world is also your structure. With your body you belong to the material realm of reality and you have physical needs. With your mind you belong to the nonmaterial realm and you have spiritual needs. Both needs require your attention. Many people think that if they take care of their body, the job is done. As it turns out, that isn't the case. If you disregard the spiritual needs of your mind, you will soon find out that your body will get sick.

The second aspect of reality that is important for our way of life is the fact that the universe is an indivisible wholeness. In a world in which all things and people are interconnected, a different way of life is needed than in a mechanical system of disconnected things. In a coherent world, everything that supports the coherence is good; everything that disrupts it is bad.

It doesn't make any sense to cheat, lie, hurt, and steal, because ultimately you are doing this to yourself. We can give something away or help another with joy, as though we were receiving or being helped! In your love of others, you are being loved. In the teaching of the Indian sages, the concept of *Karma* means that the way you treat others determines your personal fate. What you bring upon others will ultimately fall back on you!

Sociobiologists believe that pleasurable behaviors are adaptations. If a behavior is an adaptation, it means that your ancestors were selected for it. Crudely put, those of your ancestors who pretended to be helpful and kind got all the girls! Thus, according to sociobiology, the joy that we feel in helping others is a trick of our genes.

In contrast to this I propose that moral acts are pleasurable because they bring us in contact with the One. This is the biblical story of the eviction from paradise. Spat out into the empirical world, we have a longing—an inner need—to return. This need is the basis of our moral instincts because the traditional virtues, such as love, honesty, kindness, and sincerity, are connecting values. When we act with kindness, we feel the joy of getting in contact with the One. Vice versa, living an evil life will make us restless, because it destroys the connection. An evil person, Plato says in his *Gorgias,* "leads a robber's life." Such a person "is incapable of communion" and "incapable of friendship." But communion and friendship, Plato points out, are important for our life because they "bind together heaven and earth and gods and man."

"I do not know what your destiny will be," Albert Schweitzer, the twentieth-century Alsatian philanthropist, is reported to have told one of his audiences, "but one thing I know: The only ones among you who will truly be happy are those who will have sought and found how to serve." In this way we find in the depths of the quantum reality suggestions of virtues that are identical with the virtues of some of the great moral minds of our history.

The connection between virtue and happiness has been a recurring motif in practically all of the great moral systems of our history. Happiness plays an important role, for example, in Buddhist teaching. (See the inspiring books by the Dalai Lama and the French molecular biologist and Buddhist monk Matthieu Ricard.) To Matthieu Ricard, happiness is a "skill" that you have to develop. As always, a skill belongs to the potential that you must actualize.

Aristotle believed that the virtuous life will lead us to the highest good that human beings can achieve. His term for the highest good: *eudaimonia.* The philosopher Michael Hauskeller has pointed out that, whereas *eudaimonia* is often translated as "happiness" or "blissfulness," it also has the specific meaning of a happiness that follows from the "fulfillment of a successful life." You won't be amazed that *successful* in this context doesn't mean how high you climbed the corporate ladder or how much money is in your checking account!

In the philosophy of Epicurus, the fourth-century-BCE Greek philosopher, the concept of *ataraxia* essentially means peace of mind. *Ataraxia* is a state of tranquillity that you find in the absence of distress. When you are free of fear or anxiety, you can find peace of mind. "Pleasure signals to a sentient being what is beneficial for it and its constitution," writes Michael Hauskeller, whereas "displeasure and pain indicate what is harmful." An important way to find peace of mind, Epicurus taught, is in social bonding. "Among everything that wisdom contributes to the happiness of life," Hirschberger cites Epicurus, nothing is more important "and joyful than friendship."

In the metaphysics of Immanuel Kant the *highest good* is an important concept. In Kant's *Critique of Practical Reason,* virtue is defined as the worthiness of being happy, and happiness is the appropriate reward for being virtuous. The highest good is the connection of virtue with happiness.

But, as Kant pointed out, there is a problem with the highest good: Virtuous people don't seem to do well in this world. More often than not it is the crooks that get rewarded, and the "sufficient connection of happiness and virtue" isn't a fact of life that can be counted on. To deal with this difficulty, Kant introduced God as the power that guarantees the highest good.

In Kant's approach, the need of the highest good is primary, while the assumption of God is secondary. In our approach—that of Valadas Ponte, Roy, and myself—the cosmic spirit is primary, while the highest good comes out of it. Because the cosmic spirit is in you, virtuous acts will always and instantly contribute to your feeling of wellness in *this* life.

So, what should it be? The life of the fittest in an aggressive world? Or the peaceful life in the wholeness of the One? Are these merely two different options, equally acceptable, and you can pick the one that best fits your taste? Contemporary neurobiology says no. The life of the fittest is in conflict with your physiology!

In his book *The Biology of Belief,* Bruce Lipton describes how a constant state of aggression will lead to somatic stress. Extended periods of stress, in turn, can inhibit life-sustaining processes and make you sick.

According to recent neurobiological studies, human beings aren't laid out for conflict and aggression, but for social bonding and successful personal relations. You will find an excellent description in Joachim Bauer's book on human kindness. Centered in our brain is a neurobiological motivational system that rewards us for charity and kindness. As Bauer describes it: "To find and to give interpersonal recognition, respect, care, or affection is the core of all motivation." Since your brain "turns psychology into biology," human beings who live in accordance with the spirit of the One have a more wholesome chemistry than those who live in the stress of uninterrupted readiness to attack.

The true nature of reality isn't found on its surface. Peace of mind isn't found in the mindless satisfaction of bodily needs, but only in the interactions with the One.

At numerous points in our discussions in this book, we have

touched on the amazing synchronicity of the recent cultural developments. Here is another example: At a time when we begin to suspect that a cosmic ethic suggests a life with love and kindness in the spirit of the One, neurobiologists discover that our nervous system rewards us for love and kindness. It is difficult to think that the simultaneity of these discoveries is the sign of a random process. What does it take to convince us that we are the recipients of messages that come to us from out of the depths of the universe?

Human beings can act with true altruism, pursuing projects that are to their disadvantage, simply because they have an intuition that this is the right thing to do. In this way morality is akin to creativity. In a creative act, there is a joy of doing something that is meaningful without having any material advantage in mind. Vincent van Gogh created his paintings with a passion and dedication, even though he sold only a single painting in his entire life, and that was to his brother. In contrast, whenever people team up to do something for no reason other than making a profit, the result isn't worth any more than the money that they will get.

Connecting morality with the joy of creativity takes us into the realm of your inner potential. Actualizing a potential is a joyful experience because, in a creative act, you are getting in touch with the One. There is something intrinsically moral about the needs of your inner potential to actualize in creative acts, and inner joy is the sign of successfully responding to a cosmic task. Likewise, there is something intrinsically creative about altruistic acts.

The principles of your mind appear out of the cosmic realm of forms in your consciousness. You can't live in peace with your own mind if you are at war with the principles of the universe. Moral rules are rules of wellness; it is a matter of prudence to respect them, and ill advised to shrug them off.

Maine de Biran and the Principle of the Inner Sense of Existence

At the beginning of the nineteenth century the French philosopher Marie-François-Pierre Gonthier Maine de Biran, known as Maine de Biran, developed the notion that we can understand complex principles of the world by identifying with them in our "inner sense of existence." For example, I can understand what a force is by becoming a force, and *being* it. Or, as Maine de Biran expressed his experience, I can understand a force when I feel it "as identified with *me,* or inseparable from my existence as I sense and perceive it." In this way, Maine de Biran proposed, the meaning of important principles is revealed to us in "the fact of existence as it manifests itself in our inner sense."

Maine de Biran's concept is easily generalized and applied to all kinds of principles of our existence. Deep inside your soul you can feel, for example, the identity with a moral order that gives meaning to your life. If you feel it as identified with yourself and inseparable from your existence, it means that your life has a meaning. At this point we can't determine what the source of this feeling is. But if its source is cosmic we can count on it to be true. In that case it is in you and it is in me that the universe is finding its meaning. *You* are the meaning of the universe. To understand what the meaning is and to apply this understanding to its value-best state is your cosmic task; it is a basis of your morality and human dignity.

The meaning of the universe lies in its inner potential, and not on its mechanical surface. It is difficult to avoid the impression that we have, somehow, a key function in this process in which the universe is expressing its meaning in the world. Thus, to develop the cosmic potential in us to its value-best state is a moral task and the basis of a virtuous life.

The inner sense of existence is also the awareness of the wholeness in me, which I feel is identified with the wholeness of the universe. In the cosmic field, each one of us is a singular point in which the universe is in action.

JUNG'S CONCEPT OF PSYCHIC ENERGY

When you accept that a moral principle is identified with you, you give it power over your actions. Your actions take place in the empirical world and require physical energy. Thus, moral images, which have in themselves nothing to do with matter or energy, have the power to release physical energy into the empirical world.

In this context it is useful to consider Carl Jung's definition of *psychic energy*. Psychology, Jung says, has its own concept of energy. It has no mass or motion involved with it, like the energetic processes of physics, and refers to *the activity* (Greek: *energeia*) *of the psyche*. Even though it is used as an analogy, the concept of physical energy originally developed from the concept of psychic energy. In physics, where there is energy, there are forces. If we can speak of psychic energy, we can also speak of psychic forces: A psychic force can move your psyche.

If it is permissible to use the concept of energy in both our physics and our psychology, then it should also be meaningful to ask whether the same concepts apply to the universe, too: That is, is there a cosmic psychic energy field together with a field of physical energy? The definition of the nonphysical cosmic energy should be analogous to Jung's definition of psychic energy: That is, it refers to *the activity of the cosmic potentiality*. If such a field exists, then we should expect that it can affect the physical processes of the world in the same way in which our mind can affect our body. Actually, the conclusion is trivial because the cosmic field *is* doing just that: Through our minds it is affecting the material world. Since a willful effort that you express in the world must be preceded by a willful effort in your mind, the mental process is primary; the physical force, secondary.

The Cosmic Nature of Morality

Much of what our morality is about is found in the connection of our mind with the mindlike background of the universe. When moral decisions have to be made, suggestions appear out of this connection, telling us the right way to act. Your inner sense of existence is ultimately the sense of the cosmic consciousness in you, which you feel as identified with you or inseparable from your existence.

Since you belong to the unity of the world, the advice about how you should act appears in your mind in the form of recommendations of how you can act in the spirit of the One. In archaic times the recommendations appeared in the form of commandments; at the current stage of our development, they are subtle, not intimidating. They don't force you to do anything; they don't come to you with threats of hell and brimstone. They allow you freedom of choice, but if you choose right, there is a quiet feeling of satisfaction, joy, and happiness: of something you have done right. These are signs that, in a difficult situation, you made the right choice. Alternatively, wrong decisions are signaled by stress and restlessness.

At the cosmic level of morality, you won't find a detailed code of law with precisely outlined paragraphs like in a catechism, but a single principle applicable to all situations: *to act in the spirit of the One*. In this principle the traditional virtues appear quite naturally and spontaneously. Love, kindness, honesty, and sincerity—all are expressions of the spirit of the One.

Wherever you probe the basis of ethics, your inner sense of the fact of your existence comes to the fore. At its simplest level you will identify with the joy that you feel when you are helping others. At more advanced levels the identification with the cosmic spirit will drive your actions.

THE INNER SENSE OF EXISTENCE AS THE NECESSARY CONDITION OF RESPONSIBILITY

Personal identity is an essential condition for moral responsibility. If I am not the same person who I was yesterday and will be tomorrow, I can knock you on your head right now, take your wallet, have a nice dinner, and tomorrow be another person who has nothing to do with today's crimes. We can call this the principle of "self-permanence." Without self-permanence you wouldn't have a personal identity. Without a personal identity you wouldn't have any moral responsibility.

Self-permanence is a special case of Jean Piaget's object permanence and a necessary condition for us to have a personal identity. The question is, what is your personal identity? Is it the uninterrupted existence of your body? If not your body, then what? Something has to be there all the time as the basis of your personal identity. But how can you prove the uninterrupted existence of anything?

As far as your body is concerned, I must disappoint you. Not only can't you observe its atoms and molecules without interruption, but, in addition, these ETs in you are constantly blinking in and out of the empirical world. In chapter 2 we saw that ETs constantly vanish from the empirical world and enter the realm of potentiality when they are on their own. In the noisy environment inside your body, its atoms and molecules don't vanish in the realm of potentiality for extended times, but for fleeting moments, here and there, they can't help blinking in and out of the empirical world.

The atoms and molecules in your body are constantly undergoing transitions between different quantum states. These are transitions between states of different modes of motion: states of vibrational, rotational, and translational motion. In state transitions molecules always have to go through a superposition of states, in which they exist in a nonempirical mixture of virtual states, including the starting state and the final state. These fleeting superpositions don't belong to the empirical world but to the realm of potentiality. At the molecular level, the particles of your body are involved in a never-ending frantic dance: out of the actuality and into the potentiality; out of potentiality and into actuality; and in and out, and out and in.

The existence in the form of matter isn't an eternal state of ETs. What is lasting is the mindlike basis. "Strictly speaking," physicist Hans-Peter Dürr says, "there are no objects which are temporally identical with themselves." Thus, the permanence of your body is an illusion. It is a matter of averages. On the average, your molecules are there for you to rely on. However, "there is no being," Dürr explains, "there is nothing that exists. There is only metamorphosis, change, operations, processes." Your body seems to be there all the time, because its molecules don't all vanish from the empirical world in a concerted motion at the same time. If they would do that, you would indeed blink in and out of the empirical world: Now you are here, now you aren't.

If there are no objects that are constantly identical with themselves, how then can you have a feeling of personal identity?

Apart from such fancy quantum gymnastics, your body isn't what it was some time ago, because, as a matter of maintenance and repair, nearly all of its cells die at some time and are constantly replaced by new ones. In addition, there is no molecule in you that was there when you were growing up! The body that you are now isn't the body that you were some time ago. As a matter of fact, when you look at old photos of yourself, you don't even look the same.

So here we have a problem. To act with moral responsibility, you need a personal identity. But you can have a personal identity only if something lasting exists in you without interruption. Since your body isn't a lasting thing, who are you?

To answer this question, we have to conclude that your personal identity must be a property of your mind, since it isn't a property of your body. Inside you, you feel the permanence of your existence, and you feel it as identified with *you* or inseparable from your existence as you sense and perceive it. Maine de Biran has given us this wonderful principle, and you wonder why it isn't as famous, or more so, than *I think therefore I am.*

From this analysis it follows that morality is rooted in nature, but not in the empirical part of nature. Like our mind, it is rooted in the mindlike realm of the cosmic potentiality. Similar conclusions apply to other conditions of our morality; for example, they apply to your free will. Free will isn't found in material systems that follow deterministic laws, but it is found in the processes of your mind.

IMMANUEL KANT AND THE QUANTUM VIEW OF ETHICS

The view of morality that we considered above is similar to that of the eighteenth-century idealist philosopher Immanuel Kant. Kant proposed that the world is twofold. There is the realm of things, as they exist in themselves, and the world of their appearances in our experience of the world. The things in themselves, which Kant called *noumena,* are unknowable. We don't have any experience of things but only of our interactions with things, and the two aren't the same. You might say that we don't experience things, but our experiences of things. Specifically, Kant thought, the noumena and their appearances differ in that the principle of causality applies to the appearances but not to the things in themselves. This is so, Kant thought, because causality isn't a principle of nature, but a principle used by our mind to put our observations in order. It is a *form* of our understanding. The laws of physics, Kant concluded, aren't laws of the world as it is in itself; they are made by the human mind. In the world of appearances, causality holds strictly; in the world of the noumena it doesn't apply.

On this basis Kant was able to place our morality in the non-empirical noumenal world: In the world of the noumena, free will and moral decisions are possible. In the empirical world, they are impossible because empirical things follow causes, they don't make decisions. Since human beings are part of both the nonempirical world of noumena and the world of appearances our body is subject to the causality of nature, while our mind has a free will. As I like to say it, the roots of responsibility and moral values are in the transcendent realm of the cosmic potentiality.

I am sure you have recognized how similar Kant's philosophy is to the worldview of quantum physics. The world of the noumena is transempirical. It is the realm of potentiality out of which appearances can actualize. Kant's choice of words is interesting, too: The term *noumenon* derives from *nous,* the Greek word for mind. Thus, the nonempirical background of the visible world is mind-like.

Physiological studies of our brain don't reveal our free will. Freedom isn't a principle of the empirical world. Scientists who don't believe that a nonempirical part of the world exists typically deny that we have a free will. Contemporary neurologists, for example, propose that the free will is an illusion. But it isn't an illusion. Your inner sense of existence tells you that you have a choice. You have a free will and it is identified with you.

Freedom belongs to the potentiality of your mind. When you choose to live exclusively in the material world, denying the transcendent realm, you have no freedom: you are living the life of a slave.

This is why it is so important for you to be aware of your inner potential and its cosmic roots: Freedom, personal identity, and morality—all are gifts of the cosmic potentiality. But they are gifts only if we accept them. To develop in the best possible way the potentiality in you, in the sense of *aretē,* is a cosmic task.

We should not leave the discussion of our free will without mentioning Carl Jung's warning that our freedom isn't unlimited. In a trivial way, there are external limits set by the physical conditions of the world. For example, even if you want to, you can't fly. But in addition to such external limitations, there are internal limitations imposed on your mind by your unconscious. Jung called the mostly unconscious part of our psyche the *self,* and our consciousness the *ego.* Inside of our consciousness, as Jung describes it, the ego has a free will, defined as "the subjective feeling of freedom." (The "subjective feeling" sounds like the inner sense of existence.) However, Jung writes, "Just as our free will clashes with necessity in the outside world, so also it finds its limits outside the field of consciousness in the subjective inner world, where it comes into conflict with the facts of the self."

Even though you have a free will, you can't jump off the roof of a tall building and expect to live happily ever after. There are physical limits of what you can do. In the same way, Jung points out, your unconscious limits your freedom by affecting your consciousness "like an *objective occurrence* which free will can do very little to alter."

HUMAN MACHINES AS AMORAL ROBOTS

Related to the passion of taking things apart is the passion of the classical sciences to turn human beings into machines. If you took your body apart, you would find all kinds of structures and fluids, and molecules and atoms at a deeper level, but you wouldn't find any mind. From such considerations arose the view that human beings are machines. Since mind isn't found in its parts, mind must be a mechanical state.

The French philosopher René Descartes is often called the founder of modern philosophy. Modern philosophy is characterized by the fact that it considers human beings as central, while God and the universe are secondary. This is in contrast to the European Middle Ages, when God was primary. In the seventeenth century Descartes invented the duality of mind and matter. In the Cartesian dualism, mind and matter are some kind of substance that he called *thinking substance* and *extended substance.* You will often see the Latin expressions he used: *res cogitans* and *res extensa.*

Descartes's philosophy is dualistic, because mind and matter exist separate from each other and don't interact with each other. Since mind has nothing to do with matter, material systems, like our bodies, must be machines. All animals and human bodies, Descartes thought, are machines.

In his exciting book *Does God Exist?* Hans Küng describes how European thinking evolved from Descartes's dualism and Newton's physicism to the materialism of the eighteenth and nineteenth centuries. Robert Boyle, a seventeenth-century chemist whom students of chemistry remember for his gas law, was the inventor of the term *materialism.* In the beginning, materialism wasn't atheistic; Robert Boyle was a religious man.

But, as a material system, the universe, too, soon became a machine. And from that point on, it was only a small step to *atheistic materialism,* which Julien Offroy de La Mettrie, an eighteenth-century French physician and philosopher, introduced. In his book *Man as a Machine,* La Mettrie describes how everything human— our consciousness, thinking, and all the aspects of our soul—must be explained in terms of the structures of our body.

You may recall David Hume's claim that the physical order of the universe has nothing to do with human moral order. Nevertheless, the materialistic worldview was soon used as a basis for ethics.

Küng describes, for example, how La Mettrie concluded that, since we are machines, morality "is the art of enjoying life." In the same way, he considered religion irrelevant because it has nothing to do with our wellness.

We find the same urge to translate aspects of the physical world into aspects of human morality in the philosophy of Paul-Henri-Dietrich, Baron d'Holbach, who continued the materialistic program after La Mettrie's death. In his book *System of Nature or of the Laws of the Physical World and the Moral World,* he declares, as Küng describes it, "matter and mind, physics and ethics, as identical and religion as harmful," so that "priests must be replaced by doctors."

If human beings are machines—if our love and consciousness and all our mental phenomena are nothing but the excretions of specialized material structures—the feeling of moral responsibility is an illusion: *Human machines are amoral robots.* Here, *amoral* doesn't mean *immoral;* it denotes freedom from moral responsibility in whatever you do.

Contemporary neurology operates on the view that the brain is a machine, and our mind is one of its functions. Rodolfo Llinás is a prominent brain scientist. In his fascinating book *I of the Vortex,* he describes the "monist's perspective" of the human mind. The monist perspective is the view that the "brain and the mind are inseparable events." The brain is "a living entity that generates well-defined electrical activity," Llinás writes. This means that "this activity is the mind. . . . Mindness coincides with functional brain states."

In contemporary brain science, the monist perspective has led to valuable insights into the functioning of the brain. For example, we owe research performed within this paradigm the discovery that *the modeling of reality* is an important function of the brain. We understand the external world primarily not by the signals that our brain receives from our senses. Rather, the signals activate models that already exist in the neuronal structures of our brain. It is the experience of these models that shows us what the world is like.

The modeling function of the brain allows the conclusion that, as Llinás explains: "The only reality that exists for us is already a virtual one—we are dreaming machines by nature." An important part of this dream is the feeling of the "self" or "I" as an actually existing entity. But, as Llinás explains, that feeling is an illusion. "Self" or "I" isn't a "tangible thing," just a "useful construct." You can compare the concept of the "self" or "I" to the concept of "Uncle Sam," who isn't a really existing person. In this way the feeling of your personal identity and, we must add, the inner sense of existence, is what Llinás calls a "particular mental state. . . . It exists only as a calculated entity."

We must, of course, take seriously these claims and consider the possibility that our moral responsibility, if it is based on the inner sense of existence, is an illusion. "Self," Llinás writes, "is the invention of an intrinsic central nervous system semantic. It exists inside the closed system of the central nervous system as an attractor, a vortex without true existence other than as the common impetus of otherwise unrelated parts."

In contrast to these views we have to consider the possibility that the brain, like all material structures, is open to influences from a nonempirical part of the world. In that case, the brain doesn't have to invent or construct the "self" or "I," but it is merely a tool or instrument that can bring to consciousness whatever it is that exists in the nonempirical realm of the universe. If the roots of "self" or "I" are in the transcendent background, it isn't amazing that they aren't found inside the material brain.

"So, now we have a wondrous biological 'machine,'" Llinás writes, "that is intrinsically capable of the global oscillatory patterns that literally *are* our thoughts, perceptions, dreams—the self and self-awareness."

The view that the "I" is an illusion is the neurological equivalent of the sociobiological view that our morality is an illusion. It all fits together. Living systems whose self is an illusion must be amoral. Human machines are amoral robots! They are beyond considerations of values or dignity. If moral laws are illusions, forget about freedom, Aristotelian eudaimonia, Kant's highest good, and your personal responsibility.

It is possible that the system is self-fulfilling: If you believe that you are a machine, then that is what you will be, because that is where the actualization of your potential will stop. On the other hand, if you believe that you are more than a machine and search your inner potential for signs of transcendence, you might be surprised by what you will find.

There may have been a time in our history when we had a choice: Do we accept a transcendent part of the world, or do we reject it? I think that the quantum phenomena have led us to the point where we don't have a choice anymore: There is no denying that a transcendent part of reality exists. It makes a life with values possible and necessary, and it makes it impossible to think that we are nothing but robots condemned to the boring life of machines.

THE STRUCTURE OF MORALITY THAT THE QUANTUM PHENOMENA SUGGEST

At the highest level of engagement, the inner sense of existence signals the presence of the cosmic spirit, who is active in you. The cosmic spirit is the source of your feeling of moral freedom. In moments of distress, when difficult decisions must be made, the cosmic connection is the source of what Diogo Valadas Ponte, Sisir Roy, and I have called *tacit advice,* which shows you how to act in the spirit of the One.

People have often wondered where our moral instincts come from. In our history you can find the most disparate views, ranging from the recent claims of the sociobiologists that morality is an illusion to the claims of religions that moral laws are divine. As a solution to this conflict I propose that our moral instincts have cosmic roots; that is, they derive from the connection of our mind with the mindlike background of the universe. Moral instincts and their principles are in us because the potential in us is an expression of the cosmic potentiality.

Carl Jung was interested in this question from the point of view of psychology. "We do not know," he wrote, "where the roots of the feeling of moral freedom lie; and yet they exist no less surely than the instincts, which are felt as compelling forces."

Immanuel Kant saw the roots of morality in our reason: "The awareness of the moral law is a fact of reason," he wrote. And in his *Groundwork of the Metaphysics of Morals,* we find: "All moral concepts have their seat and origin in reason completely a priori." As always in Kant's philosophy, *a priori* here means prior to any experience. If you consider that the inner sense of the fact of existence is a principle of your mind, then the theory of ethics outlined in this chapter isn't too far away from what Kant believed it to be.

Kant's view that moral principles "cannot be abstracted from any empirical, and therefore merely contingent, knowledge" may at first sight seem to conflict with the view that the principles of ethics reflect the order of the universe. However, the order to which we are referring here isn't the order of the empirical world, but the order of the nonempirical background of reality. Thus, we can see Kant's a priori principles as actualizations of forms of the cosmic potentiality.

The most characteristic aspect of the system of ethics to which we are led by the nature of the quantum world is its reliance on a *single* principle: *Morality is the manifestation of a transempirical, tacit moral form that exists in the realm of potentiality and appears spontaneously in our consciousness when it is needed, offering its advice to our judgment and free will.*

The moral form is tacit because it belongs to the realm of potentiality. The tacit moral principle comes out of the wholeness of the cosmic realm of forms, where it must have an actually existing representation. We could call this representation *the tacit moral form,* meaning a form in the cosmic potentiality. It is exactly its tacit nature that makes it generally effective. Explicit catalogs of rules or catechisms aren't like that. They offer finely detailed descriptions with paragraphs and subparagraphs to tell you how to go wrong. When I grew up, there was in my father's house a Bible printed in 1640. It had the most fascinating descriptions of sins that you could think of, like a handbook or instruction manual. I couldn't get enough of reading about these wonderful possibilities and was eager to try them out.

The tacit moral form is primary. All explicit formulations of moral laws are secondary. They are necessarily deficient because all translations of the tacit moral form are imperfect. Explicit prescriptions represent the mechanistic approach to virtue. Use paragraph 2.1 to do this; then follow paragraph 3.2 and do that! Then, if you obey paragraph 7.3.α, you are out of trouble.

But reality isn't like this, and the cosmic spirit is no mechanism. In difficult situations you don't have the time to look for the right paragraph that tells you how to act. Furthermore, explicit formulations of moral laws can lead to situations that represent what the Greek philosophers called an *aporia*. An aporia denotes a pathlessness. Moral pathlessness exists in situations in which every possible action will violate an explicit rule. The tacit moral form isn't limited in this way, because it isn't limited by words. In the most varied situations its advice will always be a variation of its main theme— that is, how you can act in such a way that the cosmic spirit is in your actions.

It is important to be aware that the tacit moral form exists because when it appears in you it does so in a subtle way, not with fanfares and sirens. You have to be attentive to recognize its message, because it won't scream at you, and it won't claim authority by way of threats and intimidations. It will simply convey to you its tacit message, to take or leave.

You can find similarities with these views in historic systems of ethics. In his *Protagoras,* Plato considered that there is really only one virtue, so that all the virtues that appear with different names are one and the same single virtue. Immanuel Kant also believed in a single moral principle, which he called the *categorical imperative.* At the same time, our view is different because it assumes that the tacit moral form isn't only an abstract logical principle, but an actual existing entity—that is, a truly existing form in the cosmic potentiality. C. N. Villars's description of potentiality waves can be applied to describe the nature of this tacit moral form: The potentiality waves that make up the moral form should be conceived as physically real waves that exist in their own right, not merely as representations of the behavior of human beings.

Kant thought that "moral reason is out of itself practical." For us, the cosmic form in us is out of itself moral.

One of the most influential moral messages of all times, the Christian principle of acting with love, has the nature of a *form.* Interestingly, it was a physicist, Carl Friedrich von Weizsäcker, who noted this aspect of Christian love. Its effectiveness is precisely due to the fact that it is a form and not a mechanistic commandment; as a form it fits all possible actions.

The theologian Bernardin Schellenberger described the Sermon on the Mount as a symbolic act that should be compared to the declaration of the Ten Commandments by Moses. When Jesus went to the mountain, he indicated that, like Moses, he intended to speak about a law. However, Schellenberger writes, in contrast to the commandments and the implicate threats pronounced by Moses, the Sermon on the Mount "does not speak of duties, but of entirely new possibilities." The beatitudes "are no obligations—like the Commandments—to act and behave in a certain way. Rather, they open up new perspectives."

The principle *to act in the spirit of the One* is just like that. It is a form, a recommendation, and a perspective of how we can act. Perspectives define a higher level of moral commitment than commandments and their humiliating threats. By receiving recommendations, rather than threats, we arrive at a higher level of human dignity.

THE EVIL IN THE WORLD

No discussion of what can be good in the world is complete without a discussion of the evil in the world.

When we consider what is going on in the world, we see the hidden corruptors at work, leading the dance about the golden calf. Corrupt politicians, manipulative media, and ruthless businessmen: These are the foundations of our society. Countless wars in international relations, ethnic cleansing, religious intolerance and terrorism, the return of slavery, and the continued abuse of women: Evil is an unavoidable part of the world and no system of ethics will fix that.

Moral principles are archetypes in Carl Jung's sense. Archetypes are tacit forms. They have the important function of giving meaning to life, and they have the power to shape and regulate our thoughts. And herein lies a problem: Their messages depend, as Jung described it, on the state of the consciousness in which they appear. The same moral form can lead different minds to different actions, even incompatible and contradictory actions. This is a danger that we must face. The message of Jesus is a good example. In Western history it has inspired extraordinary acts of charity, but also hideous crimes. As Jung writes, much of the evil in the world "does not come from man's wickedness but from his stupidity and unconsciousness." Many examples show that this danger is real and that it exists in our time as well as in the dark ages of the past. "One has only to think of the devastating effects of Prohibition in America or of the hundred thousand autos-da-fé in Spain, which were all caused by a praiseworthy zeal to save people's souls."

The biblical story of the original sin is an allegorical transcription of how evil is rooted in the unconscious. I have always thought that it is an odd sense of justice to hold someone responsible for crimes committed by others. In fact, clan punishment was the practice of the fascists in the twentieth century.

Instead, it is possible to think that Jung's collective unconscious is a cosmic memory field that stores all the memories of humanity, good and bad. Diogo Valadas Ponte, Sisir Roy, and I have presented arguments for the thesis that the human brain can transfer forms from the cosmic field into our consciousness and from our consciousness into the cosmic field. In this way the cosmic field stores the crimes of humanity as well as the achievements. When the image of a crime appears in your mind it functions like an original sin. But when you follow its suggestions your responsibility is for your own actions, not for those of some archaic forebears who had yet to develop a consciousness. You will find detailed arguments for the existence of a cosmic memory field in the books by Hans-Peter Dürr and Ervin Laszlo.

Jung's understanding of how evil arises in the world takes the process out of the archaic context of crime and punishment and divine wrath, which have lost their usefulness. There is a "menacing power that lies fettered in each of us," Jung writes, "only waiting for the magic word to release it from the spell." Lessons from history show that, if we give in to the unconscious, "everything is liable to end in destruction." This danger exists now as ever, and it is real, Jung explains, because "the unconscious may reveal itself in an unexpected way at any time." As some of our recent political leaders have shown us, Jung warns, mankind is still standing "on the brink of actions it performs itself but does not control. The whole world wants peace and the whole world prepares for war, to take but one example." To the extent that evil rules in the world, Jung writes, "mankind is powerless against mankind, and the gods, as ever, show it the ways of fate."

Our world is inevitably structured in polarities, and in many conflicts the concepts of good and evil don't apply. To deal with this difficulty, our first decision must be to abstain from violence. Abstaining from violence is the principle of *ahimsa,* taught by the Indian sages for thousands of years. If ahimsa were adopted everywhere as the basis of our life, the world might not be a paradise, but it would be an immensely better place to live in.

The importance of education in this context isn't a novel suggestion. Thousands of years ago, Socrates already taught that *virtue is knowledge.* And in Plato's *Protagoras* we find: "Nobody commits an evil act by their own free will," but "the wrong actions rest on a lack of understanding." There is no reason to tread softly about it: It is stupid to be evil.

Kant believed that, to lead a virtuous life, *good will* is important. As he writes in his *Groundwork of the Metaphysics of Morals*: "The only thing that is good without qualification or restriction is a good will." The problem is that good will isn't enough. When individuals at different stages of their development can use the same moral law for conflicting actions, good will must be guided by additional principles. I suggest that acting in the spirit of the One—the non-dual basis—is such a supporting principle.

In contemporary psychology the importance of education for the ability to lead a virtuous life is a recurring theme. For example, within the framework of his humanistic psychology, Abraham Maslow writes: "The far goal of education is to aid the person to grow to fullest humanness, to the greatest fulfillment and actualization of the highest potentials, to his greatest possible stature. In a word, it should help him to become *actually* what he deeply is *potentially*."

Contemporary psychology shows us that the development of a human being involves different levels of consciousness. If the consciousness is undeveloped, a person's understanding of virtue can even lead to evil acts. This is the deeper meaning of Zeno's principle: To develop our consciousness to the value-best state is a primary duty, because it is the basis of a virtuous life.

Jung has warned us that we can't be optimistic about changing human nature. "Although contemporary man believes that he can change himself without limit," he writes, the fact remains "that despite civilization and Christian education, he is still, morally, as much in bondage to his instincts as an animal, and can therefore fall victim at any moment to the beast within."

To this I like to add that, while we can't change human nature, the evolution of life can do just that, and the discovery of the quantum phenomena is a sign that the appearance of a new human species is imminent. The surprising aspects of the world and the way of life that they suggest require more than a learning process to cope with: It takes a change of mind that is so fundamental that a new structure of consciousness, or a metamorphosis of our consciousness, is needed. In the next chapter, we will see how the new consciousness will be integrative and, thus, intrinsically moral.

Chapter 7

YOUR EVOLVING MIND: INTEGRATIVE CONSCIOUSNESS AND A LEAP INTO A NEW HUMAN SPECIES

"For thirty years I went about searching for God. When, at the end of this time, I opened my eyes, I discovered that it was God who was searching for me."

—BÂJEWÎD BESTÂMI, NINTH CENTURY;
CIT. BUBER/SLOTERDIJK

How often have you said "I changed my mind" or "That man must be out of his mind!"? Perhaps you've said of someone that "he isn't in the right state of his mind!" So, apparently, a mind is something you can get in and out of; or it has states, like a molecule, and can jump from one state to another.

This is a fitting point to consider these statements, because the quantum phenomena we have discussed in this book are so counterintuitive that a change of mind is needed to accept them. By a change of our mind I don't mean a mere change of our opinions or convictions, but something as fundamental as a change of the structure of our consciousness: like a rewiring of our brain. Specifically, for the average Western mind it is impossible to accept the congruence of the physical, the spiritual, and the mental that we find in the quantum phenomena. My suggestion is this: *The discovery of the quantum phenomena signifies an evolutionary metamorphosis of the human consciousness, a leap of the evolution of life into a new human species.* It shows that the structure of your mind is evolving, if you allow it to evolve.

We have already talked about Jean Gebser's thesis that mutations of consciousness have been ordinary events in our history, including advances from the archaic to the magic, the mythical, and the mental structures of our mind. Gebser also envisioned that another mutation is impending, and that it will take us to the integrative structure of consciousness. I translate Gebser's "integral" as "integrative" in order to emphasize a dynamic mind that actively integrates ancient spiritual views with contemporary rational views of the world.

The new structure of consciousness is the consciousness of a nondual world, the world of the quantum phenomena, in which seemingly incompatible opposites are integrated or reconciled. The integrative mind can accept cosmic order as a model of human order. It can combine spiritual views of the world with a rational understanding of cosmic order. It can find a way to base public order on cooperation and kindness, rather than competition and conflict. And we will be able to live to the fullest our individual potential in a holistic world in which all things and people are one.

Physicists such as Bohm, Eddington, Heisenberg, and Jeans started this process. All of a sudden, they found themselves in a reality in which the invisible is real and mindfulness must be accepted as a property of the universe. With this, they were drawn into ancient structures of consciousness, but their minds didn't become archaic or magic minds because the ancient concepts appeared quite naturally, without any stress, within the rational structure of physics. The outcome is something absolutely exciting: an integrative view of the world that accepts all the virtues of the past, unifying all aspects of reality: its physical order as well as its mental and spiritual order.

Previous jumps in the evolution of life typically occurred in times of global stress, when life as a whole was challenged and survived through the cooperation and mutual support of its various forms. Thus, when unexpected aspects of the wholeness of the world appear together with signs of stress, as this is happening at the present time, it seems prudent to prepare for another movement of the evolution of life.

Our Culture Is One

In 1959, C. P. Snow published *The Two Cultures,* in which he described how the arts in Western countries are typically isolated from the sciences. As a rule, artists don't talk with scientists, and scientists have little respect for artists and people whose training is in the humanities. In the same way, American businessmen have little regard for academics. Within the physical sciences a similar phenomenon is often found in a latent animosity between theoreticians and experimentalists.

The cultural rift described by C. P. Snow is an expression of the mechanistic mind-set. It is the sign of a society that has a passion for thinking in boxes and taking things apart. In reality, our culture is one.

Underlying all the modes of our thinking is the One. The synchronicity of the cultural revolutions in the twentieth century shows their connectedness. When the One changes its state and makes a quantum jump, dislocations appear in the human coordinate system like geological faults appear during an earthquake. Therefore, science isn't disconnected from the humanities; all modes of our thinking have common roots in the cosmic realm of forms. With the discovery that such a unity exists, you have begun the integration of your consciousness.

The unity reveals itself in many examples. It appears in the connections that we find between physics and metaphysics, and in the congruence of the mental, the spiritual, and the physical. To get a feeling of the congruence, just consider your own situation. In the course of your life you have experienced the world and developed a view of its order. Can you really prevent this view from affecting your ideas of how you should live? Similarly, if you were to accept the premise that the background of the universe is mindlike, could you really keep that discovery separate from your spiritual convictions? There is no doubt that our view of the world affects our way of life. All modes of our understanding have the same cosmic roots. This is the reason why quantum physics has something to reveal about how we should live.

It is no accident that, when quantum physics discovered a realm of forms at the foundation of the visible world, Carl Jung discovered a realm of forms at the foundation of our mind. What the unconscious is to the mind, the nonempirical realm of reality is to the empirical world. As elements of our mind, the archetypes are nonempirical. Since they have the potential to appear in our consciousness, they form a realm of potentiality. I think that we should have the courage to integrate the two and think of them together: Jung's realm of forms and the realm of forms that quantum physics discovered. In the introduction we have already considered Jung's passionate description of the realm of the archetypes as a boundless wholeness in which he becomes one with all that is. When material particles become potentiality waves, they also lose their identity in a cosmic ocean that is one. Jung's description of the ocean of the unconscious is the description of a mystic. In the same sense we can say that quantum physics isn't only a form of idealism, but a form of mysticism, too.

The stuff of the world is mind-stuff, Eddington said. In the human mind, the cosmic mind-stuff appears in the form of archetypical principles or concepts. In the external world, the mind-stuff appears in material structures. The concepts in our mind are models that agree with the external world because they have the same roots as the reality that they are modeling. In a synchronistic process the same principles appear in our mind that actualize in the external world.

We have a body and a mind because the universe has a body and a mind. When you become aware of the common roots of seemingly disconnected phenomena and principles, you have advanced to an important level of integrative thinking. From there you can proceed to yet higher levels, at which you allow the integrative mode of thinking to restructure your consciousness.

Because it makes the connection, the integrative mind is a mystical mind.

THE INTEGRATIVE CONSCIOUSNESS

When the human consciousness undergoes a mutation, the effects are comparable to the appearance of a new species: It is like a new animal appears. Understanding the past structures of consciousness is relevant now, Jean Gebser taught, because they are still present in the depths of everybody's mind and affect our current thinking. It is almost as if humanity doesn't consist of a single human species; ancient structures of consciousness continue to survive and coexist together with new ones. Even today, people continue to exist in the archaic structure, while others have advanced to the mythical or mental structures. In addition, Jean Gebser thought, understanding the past will help us understand the current mutation. I think that the discovery of a new world in the quantum phenomena is a sign that such a mutation is, indeed, happening to us.

Each level of consciousness, as Gebser describes it, is characterized by its specific views of space and time. Quite generally, our understanding of space and time is important for us because it defines our place in the world and affects our awareness of things. Just think about your own reactions when you discovered that the visible world is only the surface of a deeper and hidden space.

Structures of consciousness, Gebser explains, can be described only with hindsight, since an evolving structure is reached in an unpredictable "jump." You can't go wrong when you call this a *quantum jump*. It is like the cosmic consciousness takes a leap and makes a state transition in us! As a consequence, in the appearance of the new consciousness our mind is taking a leap. As in the quantum jumps of atoms and molecules, you can't be sure of the state on which the jump will land. Nevertheless, Gebser thought, there are signs that the currently forming structure will "reconstruct the human 'factor' from its parts in such a way that it can willfully integrate itself with the Whole."

The last statement seems to imply that, in the current structure—the mind of the Classical sciences—the human factor has been broken into pieces and it will now be healed by its integration with the Whole. Is it a coincidence that Gebser appealed to the "Whole" at about the same time the principle of wholeness emerged in a completely different context in quantum physics?

Gebser also called this ongoing process the "restoration of the unharmed, original state." The original state was the archaic structure of consciousness, in which human beings felt one with the world and "were not distinguished from the universe." We will, of course, not restore the archaic state in its original form, but we will integrate those of its aspects that we accept as true with our current rational understanding of the world.

Previous structures of consciousness typically considered their predecessors as primitive. The mind-set of the classical sciences, in particular, looks down on anyone who finds truth in nonempirical principles. In contrast, in the newly emerging integrative consciousness, archaic, magical, and mythical motifs emerge naturally in a rational understanding of the world.

The metamorphosis that started in 1900 is still turning the world upside down. Our time is a time of upheavals: upheaval of our view of the world, upheaval of our view of humanity, and upheaval of our view of God. Riding the waves of the cosmic ocean, we can't tell where the storm is blowing us. But one thing is clear: the direction, if we are doing it right, will lead us away from aggression and selfishness to a world of human kindness and bonding.

Alternative Medicine as an Example of the Integration of the Mind

Joachim Faulstich has illustrated the nature of the integrative consciousness in the context of alternative medicine. In his inspiring book *Das Geheimnis der Heilung* (The Secret of Healing), he gives the following account.

Conventional Western medicine is based on the materialistic and mechanistic worldview of the classical sciences; Gebser would have said it is based on the mental structure of consciousness. Within this structure, the human body is a machine. When a person is sick, the mechanic will analyze the symptoms and repair the machine by replacing the faulty parts. Everything else, in particular anything spiritual, is regarded as irrelevant. Doctors are engineers and businessmen, and in the United States, the health care system has become the health care industry.

These practices, no matter how helpful they are in many cases, neglect the fact that we have a mind as well as a body and that the mind can affect the body in many ways. In addition to the Western tradition, there is a second medical tradition, that of the shamans and spiritual healers, who believe that when the body is sick, it is the mind that is sick and needs attention.

The two contradictory traditions of medicine can also be characterized by the voices they obey: One follows the voice of reason, while the other follows the voice of feeling. Reason is that authority in our mind that always thinks itself superior to everything else. Thus, medical doctors of the Western tradition typically consider spiritual healers as frauds or crackpots. This is, as Faulstich explains, where Gebser's integrative consciousness can come into play: It allows us to accept the power of spirituality without abandoning the rationality of the Western sciences. It allows spiritual principles to reveal themselves in rational analyses.

Many medical facts don't fit the program of Western medicine: Spontaneous healings, the power of faith, placebo effects, the successes of psychosomatics and hypnotherapy—all these phenomena can't simply be shrugged off as freak appearances. In Jean Gebser's sense, Faulstich suggests that integrative thinking will allow us to "respect the rational component in us as well as the magic component of our personality and give both a voice." Respecting the magic mode of understanding the world isn't an attempt, Faulstich emphasizes, to return to the times of magical and mythical thinking. At the same time, he points out, we have to realize that the future of the medical sciences doesn't rest in "cruel rationality." Accepting nonrational ways of reasoning for rational reasons is integrative thinking at its best: "For rational reasons," Faulstich explains, "I propose a cautious opening to the nonrational, to a plane of the soul, which lives primarily in images and stories, or in what I call 'the magical.'"

In a holistic world only an integrative approach to life makes sense. "Pioneers of an all inclusive medicine," Faulstich says, are searching for alternative ways of healing, in which "body, mind and soul return to a state of equilibrium." We might add that this state of equilibrium should be searched for in all aspects of life: in politics as well as in business; in our educational practice as well as in our social order; and, particularly, in our daily life.

How the Universe Reveals Its Meaning in Its Integrative Nature

Does the universe have a meaning, or is it a meaningless process that runs its course without any purpose? That is an important question to ask, because we belong to this universe. If it is meaningless, our own existence is meaningless. Since we don't see any purpose in the phenomena of the visible world, scientists since the European Renaissance have insisted that there is none. In this way the universe became a machine, rumbling and rattling mindlessly along meaningless tracks, going nowhere. If you don't see it, it doesn't exist!

In this context it is important to realize that the character of the universe is integrative. This means that physical lawfulness and spiritual meaningfulness don't exclude each other, but can coexist in a single system. Just look at yourself: You can adopt a purpose and give your life a meaning, even though your body follows the laws of chemistry. This is possible because a meaningful life is a part of your inner potential and transcendent being. It doesn't interfere with the functions of your body in the external world. As soon as you find a meaning for your life, the universe isn't free of meaning any longer and, you might say, it has found a meaning. The cosmic processes, too, can be lawful and meaningful at the same time: If nowhere else, the universe finds its meaning in you.

Jacques Monod was amazed by what he described as the *paradox of biology*. This is the paradox that, on the one hand, the scientific description of nature must be objective; that is, Monod said, it must exclude any reference to purpose. On the other hand, the biological sciences force us to admit that living systems are embodiments of purpose.

I think that Monod saw a problem in this, because he equated purpose with bias. By this I mean that he seems to have thought that if processes follow a purpose, they aren't entirely lawful because the end justifies the means. In that case, the laws can be bent if it is good for the purpose. Like fire and water, Monod thought, lawfulness and purposefulness don't mix. What Monod didn't consider was the possibility that the universe can be lawful in its empirical realm and, at the same time, purposeful in its nonempirical realm. In the empirical world, the laws are objective and absolute; in the nonempirical realm other factors can come into play. You know it from yourself: You are subject to the laws of nature but, deep inside you, you have the potential to find a meaningful life. Scientists don't find purpose in the measurable world. But that doesn't mean that all of reality is free of purpose. There is a connection here with the question of creativity. In a deterministic Newtonian world creativity is impossible. In such a world the future is determined by the present and completely predictable. We could say that the cosmic potentiality in such a world is closed. In contrast, in the mindlike background of the universe that we have described, causality doesn't seem to apply. Thus, the background of the visible world can be creative, like the human mind. The fact that we can be creative—creating the unexpected and unpredictable—already demonstrates that something is wrong with Newton's universe and Monod's paradox.

Deep inside its potentiality the universe is a message that wants to be spoken, and it wants you to speak it. Being a meaningful message is an essential character of the cosmic potentiality. That message isn't some random noise; you can look at it like a poem. What sense would it make for a cosmic system to develop a realm of potentiality if it didn't want to express it in a meaningful way? And it is in you that the cosmic potentiality finds and expresses its meaning. I am not sure that we will ever understand what that meaning is, but I am sure that we won't understand it by making experiments and measurements of the external world. Instead, we have to search for the meaning deep inside our soul. By *soul* I mean the part of our psyche through which the cosmic potentiality communicates with our mind.

It has often been said that, whether you like Jacques Monod's philosophy or not, you can't help being excited by his book *Chance and Necessity*. It is a passionate confession, and even where he goes wrong, Monod helps to clarify important issues that have been misunderstood for a long time. Like Jean Gebser, Monod believed that the process of evolution hasn't only affected our body, but also our mind. But he was thinking more of a process of cultural evolution than a change of the structure of our consciousness.

Monod was inspired to his views by our cultural history. For human communities, he pointed out, maintaining the cohesion of the group is a challenging problem because we aren't like ants and bees, whose social behavior is genetically programmed. According to Monod, to solve this problem, traditional societies resorted to religious stories and myths that told people what the world was about and, therefore, how they had to act. The myths typically claimed that the universe has a purpose and that, therefore, our life has a meaning. Monod calls this the "animist tradition" of our history. Its main characteristic and the basis of its strength was the idea that knowledge and values were derived from a single source and human beings were at home in the universe.

In this world of myths, the kind of science that evolved in the age of the European Renaissance led to a crisis, because it denies any connection between moral order and cosmic order. If you were asked to name the single most important achievement, or most characteristic aspect of contemporary science, what would you choose? For Monod the answer was easy: The most important achievement is the discovery that nature is objective and that the principle of objectivity is "the only source of real truth."

When processes occur for no reason, they are meaningless. A mechanical universe is a meaningless machine. Everything in it is meaningless and, yes, that includes you. And with this we have touched on a number of problems that show that the principle of objectivity in science, as Monod understood it, isn't as harmless as you might think at first sight.

The first problem with what Monod calls the *objectivity of nature* is that human beings have a craving for meaning in life. Think about yourself: Do you have the feeling that your existence is meaningless? Or when you think of your family and friends who depend on you and love you, do you think that your life has a meaning? If both possibilities were an option, which one would you prefer: to lead a meaningless life or a life that has a meaning?

Monod believed that most people have a craving for meaning in life, because we are the descendants of animists; somehow the needs and beliefs of our ancestors have found their way into our genes. "Every living being is *also* a fossil," he writes. "Within it, all the way down to the microscopic structure of its proteins, it bears the traces, if not the stigmata of its ancestry." Thus, Monod explained, the stigmata handed down to us by our genes are the reason why so many people feel a "sickness of spirit," when they realize that their life and the universe are meaningless. "We would like to think ourselves necessary, inevitable, ordained from all eternity," he writes.

Monod saw a second problem that is caused by the objectivity of science and nature. It is what he calls an "epistemological contradiction" and arises in the following way.

When you look at the physiological processes of living organisms, you can't avoid the impression that they have a purpose. The chemistry of your body is tuned to a task. Its purpose is to keep you alive. Keeping you alive isn't their accidental effect, but their purpose. Living organisms, Monod admits, are endowed with a purpose. Their purpose is "to preserve and reproduce the structural norm." Another aspect of purpose is found in what he calls the "autonomous morphogenesis" of living organisms, meaning the self-governed development of the bodily form that is typical for a species.

So here is the logical difficulty, as Monod describes it: On the one hand, science forbids us to think that the processes of life follow a purpose. On the other hand, "objectivity nevertheless obliges us to recognize the teleonomic character of living organisms" (where "teleonomic" means "purpose-driven"). But you can't have it both ways. Either there is purpose, or there is none.

When you read Monod's account of this matter, you realize he was truly troubled by it. In fact, he was so concerned that he called this contradiction "the central problem of biology . . . which, if it is only apparent, must be resolved, or else proved to be radically insoluble." Because it is an embarrassing problem, biologists often try to polish it away. They call the teleonomy, or purposefulness, of living organisms their "apparent" teleonomy. Like, the chemical processes in your body just happen to keep you alive; they really don't mean it. However, as you will see in a few moments, Monod had it right: The epistemological contradiction *is* the central problem of contemporary biology, because it shows that its very foundation is flawed.

You could compare the embarrassment of biologists who find a purpose behind the visible surface of living beings with the embarrassment of physicists who find a nonempirical reality behind the visible surface of things. It makes sense to think that the two phenomena are connected.

We have arrived here at a rather important point of our discussion, because, with his epistemological paradox, Monod discovered a principle that is of great general significance, even beyond biology. The discovery is an expression of the fact that the empirical realm of the universe has a transcendent background. Purposefulness and mechanical lawfulness are principles of two different realms of reality. Purpose is a principle of the transcendent realm; mechanical lawfulness is a principle of the empirical world. In an integrative biology—which recognizes the power of the transcendent world on our bodies—the coexistence of objectivity and purpose is no problem at all. In the purpose of living organisms, the purpose of the universe comes to the fore.

At the moment you accept the reality of the nonempirical realm of the world and of its purpose, the integrative structure of consciousness further develops in you. If there is purpose in the realm of potentiality, it is also in you. Vice versa, the purpose that you feel in you is, quite naturally, the purpose of the universe, too.

What is good for the universe is good for life. There is no conflict in assuming that the processes of living organisms are both, lawful and purposeful. It is the same as having a body with its actuality, and a mind with its potential. We don't know what is going on in the cosmic potentiality, but we know that our own potential wants to come out. You can feel its urge inside you. This is another case of the power of the inner sense of existence. Your inner sense of existence shows you that the potential in you has a purpose, because you *are* that purpose and identify with it. Its purpose is to actualize the cosmic creativity in the empirical world.

In a more mundane manner, think about the departments of a government. In the United States, for example, there is a Department of Agriculture, a Department of Commerce, of Defense, of Education, and so on. All these departments have a purpose. Nevertheless, in their operations they follow the law. (At least we hope so!) In the same way, the universe can have a purpose and pursue it in a lawful way. That the one is right doesn't mean that the other one is wrong. This is the integrative structure of consciousness at work.

So here you have it: Living beings aren't special. Our nature is the nature of the universe. Since the purpose in us came out of the wholeness and belongs to it, the conclusion must be that purpose is a cosmic property.

The intent of integrative thinking isn't, as Joachim Faulstich emphasized, to turn the clocks back to archaic times. On the contrary, the intent is to support the metamorphosis to a new consciousness by integrating previous achievements with current insights. In the integrative mode of thinking, Monod's paradox isn't a problem but a promise.

We can't know what the purpose of the universe is, but we might make a guess: Its purpose is to take the structure of consciousness to ever increasing levels of integration.

PHYSICS OF ENLIGHTENMENT

The reality of the virtual states of atoms and molecules is a decisive argument for the existence of a nonempirical part of the world. Without their empty states, the atoms and molecules couldn't do a thing. Thus, invisible virtual states are essential for the real world.

It seems safe to assume that, when theoretical chemists became aware of the empty states of molecules, they didn't search our spiritual history to find an appropriate name. We can only guess that when they called empty states *virtual,* they were thinking of something that exists virtually, but not really. But what exactly does that mean? When you choose a name for something, you are making a statement, and more often than not, that statement is in the hidden meaning of names.

The word *virtual* has Latin roots. All kinds of Latin words come into swing in this one. *Vir,* for example, means "man," and it means it in a male chauvinist way. *Virtus* means "manliness," but also "virtue," "value," and "strength." This is probably what the medieval theologian and mystic Meister Eckhart had in mind when he invented the concept of a "virtual being." Yes, you read that right: It was a medieval mystic, not the quantum chemists, who first invented this term!

Meister Eckhart wrote wonderfully inspiring mystical texts.

The German philosopher Joachim Kopper, who worked in the second half of the last century, has written a classic book on Meister Eckhart's philosophy, *Die Metaphysik Meister Eckharts*. As he explains, Meister Eckhart writes about the universe: "In its entirety it stands in the oneness of God, which originally rests in itself." Oneness is the passion of the mystics. For Meister Eckhart it was a recurring theme. "In its foundation," Kopper describes Eckhart's view, "the universe is entirely contained in the oneness."

Now, even an inspired mystic such as Meister Eckhart must have been aware of the fact that you have some explaining to do when you place the universe in the wholeness of a hypothetical One, in spite of the fact that the world as we see it consists of countless separate things. However, for Meister Eckhart, there was no problem. "The things," Kopper describes Eckhart's view of the visible world, "are out of the oneness of the divine light." In the empirical world, the visible things "obtain their reality from the multiplicity" of all other things. "In God, however, they are in the oneness."

In the treatise in which he discusses these things, Meister Eckhart's interest is in the question of what makes things real. When you read this, you might be amazed. You would expect a mystic theologian to be interested in God, not the nature of things. And besides, we are always told that it is the privilege of scientists to tell us what the world is like. But in ancient times science and spirituality weren't as disconnected as they are today and mystics have the same passion for finding true reality as scientists; they just go about it in their own way. Of course, when he searches for the nature of reality, Meister Eckhart isn't looking for matter or energy or some force fields. Instead, he proposes, as Kopper describes it, that the "standing in the oneness and truth of God is the ground of the earthly being of the things." In this Eckhart enters a new dimension: Not some physical variables, but the "truth of God"— something nonempirical and mindlike—is the basis of reality. If you think that this is an odd way to describe the world, just ask yourself how different Eckhart's view is from Eddington's statement that the stuff of the world is mind-stuff. And if it is mind-stuff, what could be its source?

Unfortunately, Meister Eckhart thought, the things of this world don't understand what their true being is. They "can never comprehend their own ground," Kopper describes Meister Eckhart's view, "precisely because in this world they find their fulfillment in the nothingness." Meister Eckhart didn't say this, but I am tempted to add that the same principle applies to you and me: If we search for our fulfillment in the nothingness of the empirical world, we can never understand our own true nature and potential.

With this we have arrived at the theological foundations of the virtual states of quantum chemistry. "The 'esse virtuale,' " or the *virtual being* of things, Kopper cites Meister Eckhart, is their foundation in the "oneness of God's being, in which it is completely contained."

I have to confess that, every time I read this—and I have read it many times—I am stunned by the appearance of ancient mystical concepts in contemporary physics. What exactly is going on in their minds when scientists are engaged with their research? Does the appearance of mystical concepts in quantum physics mean that the virtual being of ETs also rests in "the oneness of God's being," even though the quantum physicists emphasize that they have nothing to do with such stuff?

Meister Eckhart usually chooses his words with deliberation. When he speaks of virtue, he means virtue. Specifically, he says, as Kopper explains, that the being of things in the unity of God's being is "the 'virtue' in the things." Translating it into current language, we might say that their virtual being is their true being. When things actualize "in the earthly order," as Kopper describes Eckhart's view, their "form is won from a state outside of themselves."

Thus, according to Meister Eckhart, the visible world of things is secondary and inferior; the invisible world and its wholeness are primary and virtuous. Do the same principles apply to the virtual states of atoms and molecules and their actualizations in the empirical world?

You will recognize in these views the typical attitude of mystics: When you are in your body, the mystics say, and not in the oneness of God, you are outside of yourself. Apart from any mystical experiences, our inner sense of existence, too, seems to convey the feeling that our body is secondary, while mind and soul are primary. When you begin to understand your inner sense of existence as the feeling of your being in the "oneness of God's being," it will open the door to your infinite potential.

Given this historic background, we don't have to be amazed that the virtual world of quantum theory has the nature of a potentiality with aspects of consciousness and wholeness. Many scientists will react with outrage against any attempt to bring their art into the context of a natural theology. But the fact is undeniable that, by the manner in which it describes the external world, contemporary physics has taken science into questionable company: The wholeness of quantum reality, the virtual states, reality as potentiality and actuality, the nonempirical realm of reality, the transpersonal images in our mind, and the potential in us—all these concepts have been known for thousands of years in our spiritual history. You can consider their "discovery" in physics as a sign that progress in science isn't driven by new types of experiments, but by the images in our soul.

Gerald Hüther believes that the power of the inner images in us is so absolute that it is possible to think that the historic achievements of humanity aren't really the work of historic personalities, but of the inner images active in these people. This puts the inner images in us in the context of something nonpersonal that befalls us, like a fate or destiny. In this you may also hear echoes here of Hegel, when he said that the world is ruled by the "cunning of ideas."

As a person who has spent his life as a scientist, I am fully aware of the fact that any reference to anything nonempirical and spiritual is a sacrilege in science. But I don't know how to avoid it. The inner images of psychology, neurology, and quantum theory are a fact. They aren't matter or energy—Meister Eckhart might call them "the virtue in us"—but they have the power to affect the material world. They do it in our brain, and they do it in all molecular processes.

Thus, we must think that the inner images are also at work in the evolution of life. If there is any hope for us, it is the possibility that our continuing evolution is the adaptation, not to increasingly brutal monsters, but to the nonempirical forms of the cosmic potentiality. By this I mean a process in which our potential is constantly enhanced in such a way that we can fish increasingly complex forms out of the cosmic ocean.

For thousands of years human beings have reached for the transcendent in the world: in prayers, meditation, or offering gifts to their God. Modern scientists believed that they had liberated humanity from such superstitions. That science now supports the search for the transcendent comes as a shock. It is the shock that drives the metamorphosis of the mind. Searching for the transcendent in the quantum phenomena, we are finding out that the transcendent is searching for us.

It is no accident that the description of the world offered by quantum scientists echoes Meister Eckhart's description, because mysticism is just another empirical method of exploring the world. Meister Eckhart's mysticism is a special kind of physics: the physics of enlightenment.

HOPE

In the seventeenth century, at the birth of the age of classical physics, Blaise Pascal, French philosopher, mathematician, and scientist, already sensed that the newly evolving science was isolating humanity from the rest of the world, and it filled him with horror. In his inspiring book *Does God Exist?* Hans Küng gives a gripping description of Pascal's existential distress. The following citations by Pascal are taken from Küng's book.

"The eternal silence of these infinite spaces fills me with dread," Pascal wrote. It bothered him to no end that we "have come out of nothingness and are carried onwards to infinity" in a process that doesn't make any sense to us. "I do not know who put me into the world, nor what the world is, nor what I am myself. I am terribly ignorant about everything." In the centuries to come, this ignorance was to lead humanity to an existence in a mechanical world in which no meaning could be found. "I see the terrifying spaces of the universe hemming me in," Pascal wrote, "and I find myself attached to one corner of this vast expanse without knowing why I have been put in this place rather than that."

The infinity around us and our own nothingness were the source of what Pascal called "man's wretchedness." All that he could see was "infinity on every side, hemming me in like an atom or like the shadow of a fleeting instant." In this infinity of everything, our senseless limitations and the unavoidable end that is waiting for us are an outrage: "All I know is that I must soon die, but what I know least about is this very death which I cannot evade. Just as I do not know whence I come, so I do not know whither I am going." So what is the upshot of such a life outside of the order of the universe? "The last act is bloody, however fine the rest of the play. They throw earth over your head and it is finished forever."

In contrast to Pascal's existential despair: the quantum universe.

Quantum reality offers reasons for hope. When reality is an undivided wholeness, we aren't hemmed in by its infinity, but we belong to it. When the background of the universe is mindlike, we aren't alone in the universe, but the cosmic spirit is thinking with us. In your thoughts are divine thoughts. In your kindness, divine kindness comes to the fore. And in the potential in you, the infinite divine potential is trying to express itself in the empirical world. Why it needs us, I have no idea. Perhaps the answer is that we are the cosmic spirit and the cosmic spirit is us.

BECAUSE THE CONNECTION IS MYSTIC, THE INNER POTENTIAL IN YOU IS DIVINE

James S. Cutsinger reports that, when a disciple asked Jakob Böhme, the seventeenth-century Lutheran mystic, how he could make contact with God, the master answered: "Son, when you can throw yourself into That in which no creature dwells, though it be but for a moment, then you shall hear what God says."

When the disciple asked where he could find "That" in which God could be heard and seen, and whether it was near or far away, the master answered: "It is in you, my son. If you can for a while but cease from all your own thinking and willing, you shall hear the unspeakable words of God."

You may hesitate to use the words of a mystic to describe the inner potential in you. But since it is connected with the cosmic potentiality, your inner potential is something mystic. "That in which no creature dwells" is the nonempirical realm of reality. Thus, when Böhme tells his disciple that "God hears and sees through you, being now the organ of His Spirit, and so God speaks in you, and whispers to your spirit, and your spirit hears His voice," then you can take this as a metaphor of how the potential in you is working: Through your inner potential the cosmic potentiality whispers in you. Since we are here describing the indescribable, the choice of words is irrelevant. Whatever the background is, you can call it the cosmic potentiality, the wholeness, the One, the nonempirical realm, or god—it doesn't matter.

Because the connection is mystic, your inner potential is divine. Because it is divine, it is infinite.

Martin Buber reports that when Bâjewîd Bestâmi, the ninth-century Sufi mystic, searched for God, he made a surprising discovery. "For thirty years," Bestâmi reports, "I went about searching for God. When, at the end of this time, I had opened my eyes, I discovered that it was God who was searching for me."

Searching for your inner potential is one of the most important tasks of your life, and it is like Bestâmi's search for God: It may take you many years, even decades, to find it. But when you finally open your eyes, you may discover that it was your potential that was searching for you.

THE INTEGRATIVE GOD

If you have objections against the idea of a divine presence in you, you should ask yourself whether your aversions have to do with the cosmic spirit in itself, or with the many crimes that have been committed in its name. The concept of a cosmic spirit appears in the quantum phenomena as a benign principle, a suggestion in a subtle way, and without any dark archaic threats: It appears in the form of an integrative view of God. The integrative God invites you to consider, if you like, whether the presence of the divine will make a difference for you.

You should consider the possibility that you have spiritual needs because the nature of the universe is spiritual. This is your own integrative spirituality because it is the same as the cosmic spirituality. If you look inside, you will find in you a desire to live in agreement with the cosmic order, because you are one with this order: This is your integrative morality. You won't lose any dignity by finding out that the potential in you is cosmic and that your thinking is the thinking of the cosmic spirit, because this is what you are: an integrative identity.

Behind the visible surface of things is the infinite ocean of possibility. Its waves are so beautiful and inviting. "What a wonderful world," Louis Armstrong sings. What a wonderful life, in which the playful waves in the cosmic ocean dare you, tease you, and play a game of hide-and-seek with you, all the time hoping that you will catch one and turn it into a beautiful poem, a painting, a song, or a wonderful act of human kindness.

Acknowledgments

Deepak Chopra's support of this book and his generous help in writing it have been one of the most wonderful and moving experiences of my life. He spent countless hours rewriting long passages of my text in order to teach me the art of presenting complex concepts in a generally understandable way. My gratitude is infinite.

My gratitude is equally boundless for Julia Pastore, senior editor at Crown Archetype. For many months she has made this book her cause, and she helped me, with incredible skill and kindness, to present my thoughts in an effective way. Without Julia and Deepak this book wouldn't be here.

I also acknowledge with gratitude the help of Gary Jansen and Stephanie Knapp from Crown Archetype for guiding the book through the final technical stages of editing; and Robin G. Roggio from the University of Arkansas Libraries for her incredibly skillful help in finding reference materials. Thanks are also due to Lauren Dong for the beautiful design of the interior of the book, and to Nupoor Gordon for the touching jacket design that so skillfully catches the spirit of the book.

And then there is Gabriele, the most important person in my life. Some fifty years ago she decided to accept me as the challenge of her life, and ever since she has supported me with incredible love and kindness, and encouraged me to become the person who I am today. Our children, Nathalie and Nicole, and their children—Caroline, Claire, Kate, and Schafer—make it all worthwhile.

It is a pleasure to acknowledge Joan and Eric Berman, Paul Drechsel, Cornelia van Eys, and Diogo Valadas Ponte for their helpful discussions of various technical issues that are important for this book; and António Cunha, David Dubbell, Robert Page, and Jim Young for their special encouragement. Along the same lines, I am grateful to my students of many decades for helping me to clarify my thoughts through their critical thinking and challenging questions.

Last, but not least, it is a great pleasure to thank all the wonderful friends in my life for their support. Each person in the following alphabetic list has inspired my thinking or supported my program, in one way or another, at some point in my life: Rita Andrade de Almeida, Barry Bostic, Odilon Bremer (Mr. Didi), Gary Cooper, Armandino Cunha, Carlos Cunha, Pedro Cunha, Chris Cunningham, Glenn Anthony Davis, Bill Durham, Günter van Eys, John Ewbank, Peter Fersley, Cristina Figueiredo, Collis Geren, Joana Gomes, Anatoli Ischenko, Roger Koeppe, René Krüger, Karl Georg Lösch, Belém Machado, Marta Maciel, Katharina Meyer, Sabine and Erich Paulus, Paulo Pessolato, Peter Pulay, Jorge Quinta, Miguel Rudko, Ana Maria Silva, Ana Maria Soares, and João Veiga.

How can we show our gratitude for our friends and the blessings in our life? By passing on to others what so generously we received, and by inviting the cosmic consciousness to think in us.

Appendix for Chapter 1

ON SINGLE-PARTICLE INTERFERENCE AND THE CONCEPT OF POTENTIALITY WAVES

In this appendix we will take a look at how the wavelike basis of reality was discovered and how the concept of *potentiality* arose in this context. We will do this by starting out with the general properties of waves and particles and with a phenomenon that physicists call *single-particle interference*.

ON THE PROPERTIES OF PARTICLES AND WAVES

Many of the phenomena studied by physicists involve phenomena of everyday life, such as the behavior of waves and particles. They are easy to understand because you already know them very well.

Everybody, I am sure, must have some experience with ball games, either watching a game or playing a game. Baseball, football, soccer, bowling, or tennis—you name it! The center of attention is always this delightful little thing, the ball, which moves about with a certain speed and bounces around. Basically, the properties of ordinary balls define very well what physicists mean by a *material particle*.

When you take an ordinary object and divide it into smaller and smaller parts, at some point you arrive at elementary units of matter that you cannot divide any more. That is the level of the elementary particles. As to ordinary things, they are all made up of atoms. The word *atom* derives from a Greek word meaning "indivisible."

As it turns out, atoms aren't really indivisible; they are made up of smaller elementary particles: particles called protons, neutrons, and electrons. Atoms are tiny things; you can think of them as tiny balls with a diameter of one ten-billionth of a meter.

The word *particle* derives from a Latin word that simply means "small part." You divide things; you end up with small parts. The concept of an elementary particle implies that, at the bottom of things, we find units of mass that behave like the objects that are composed of them. So, by definition, elementary particles are compact and localized lumps of matter; they are impenetrable and fill space solid, and they bounce off of one another when they collide in space, like pool balls.

As with balls, I am sure that you must have had some experience with waves. Sitting on a beach, watching the waves roll in from the infinity of the ocean, or swimming in them, is a wonderful way of enjoying life. Waves are everywhere. The sounds around you, for example, consist of waves: They are waves in the air that you can't see, but you can hear them when they strike your ears.

The best way to start thinking about waves and their properties is to realize that a wave is not a thing, but a process. A wave is not a static structure, but a dynamic appearance. Something—a medium in which a wave is propagating—has to move in the characteristic way of waves: up and down or back and forth, again and again, to produce the typical motion of a wave. Water waves swing in water; sound waves swing in air; light waves can swing on their own in a vacuum. The water molecules in the ocean have to move in circles to make waves, and they have to do that in a concerted way. Graphically, we can present the up-and-downness of a wave by a typical "wavy" line, as shown in figure 1. When several up-and-down movements are repeated—when two, three, or many mountains and valleys follow one another at regular intervals—we speak of a *wave train;* I am sure you have seen such trains of waves move across a water surface. The tops of the mountains of waves are also called their *maxima,* while the troughs or valleys are called the *minima.* The height of its maxima or the depth of its minima defines the *amplitude* of a wave (see figure 1).

There is a characteristic contrast between the properties of particles and waves that has to do with the fact that one is static, while the other is dynamic: Waves are appearances that are extended in space; particles are localized. A typical property of a wave is a *length,* called the *wavelength.* You can define it as the distance between two maxima or two minima (see figure 1). Another typical property is the number of times that the ups and downs in a wave train repeat each other in a certain time: The *frequency* tells us how often that happens in a second.

Figure 1. CHARACTERISTIC PROPERTIES OF WAVES

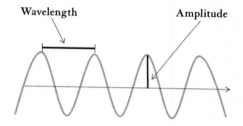

On the Interference of Waves and Wave Diffraction

What do we mean when we say that particles that collide in space bounce off of each other? We mean that, in a collision, two particles will abruptly change their direction of motion.

When two wave trains collide in space, they don't act like that. Actually, their meeting in space isn't really a collision. Waves don't bounce off of each other; in an encounter they don't change their direction of motion. Each one continues in its track, and in doing so they move right through each other. Think of two wave trains on a beach that come from different directions, roll through each other, and, for a short time, form a single train. For a while their crests and valleys superimpose, playful patterns appear, and then the performance is over: The two trains of waves reemerge out of their encounter, and each one continues its thoughtful path unharmed, as though nothing had happened. Physicists call this the *interference* of waves (see figure 2). When two wave trains interfere in such a way that crests get to lie on top of crests, they build each other up, a big wave results, and we say that the interference is *constructive;* when crests get to lie on troughs, the waves cancel, and the interference is *destructive.* So the differences in the properties and

Figure 2. INTERFERENCE: WHEN TWO WAVES MEET IN SPACE, THEY SUPERIMPOSE OR INTERFERE.

When the crests of one wave are on top of the crests of the other, the interference is constructive. The waves add: a big wave results.

When two waves meet in space in such a way that crests are on top of valleys, the interference is destructive. The waves cancel.

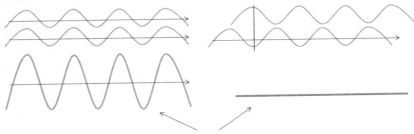

The result of the interference

When two waves meet in space, they superimpose—they *interfere* with each other: their amplitudes add and subtract. When, in that process, the crests of one get to lie on top of the crests of the other, the interference is constructive (upper left), the two waves reinforce, and a big wave results. When wave crests get to lie on top of valleys, the interference is destructive (upper right) and the waves wipe each other out. The surface of the ocean can be like an interference pattern. In some places the waves reinforce; in others they cancel. Interference is the addition and subtraction of wave amplitudes.

behaviors of waves and particles are obvious. Waves interpenetrate; particles bounce off. Waves are continuous; particles are discrete. Waves are extended in space; particles are localized. Waves can be massless, like light waves; being massy is the defining property of particles. It is obvious: One thing shouldn't have the properties of both particles and of waves.

With this, we have acquired nearly all the tools we need to define the concept of potentiality in physics. To conclude the exercise, there is just one more item to consider, which has to do with the way in which waves behave when they hit an impenetrable obstacle in their path. When waves strike on such a thing, they *bend* around its corner. Physicists call this bending the *diffraction* of waves. A simple example is shown in figure 3.

At first sight, the concept of diffraction seems a rather esoteric one, but it is quite common.

For example, have you ever wondered how it is possible that you can hear the conversation of people at the other end of a hallway while you are sitting in a classroom? You can't see these people, and the sound waves coming off them aren't directed into your room,

---

Figure 3. THE DIFFRACTION OR BENDING OF WAVES

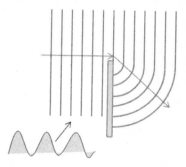

When waves in their motion through space hit on an obstacle that blocks their path, they bend around its corner. This bending is called the diffraction of waves. Diffraction is wave bending.

but you can hear them in your room because they bend around the edges of the door.

On the Internet you can find beautiful pictures of waves rolling in from the ocean and bending around the corner of some wall—for example, a harbor wall.

Another example of bending waves is observed when a light beam hits a wall with a slit in it. In that case, the beam doesn't just shine through the slit, as though nothing had happened. Its waves bend around the corners: The light is diffracted. The diffraction of light by slits has an important consequence: Behind an irradiated slit light spreads out right and left, and in all directions (see figure 4). You might describe this by saying that an irradiated slit is a source of light waves that spread out in all directions.

On Double Slits and Interference Patterns

The diffraction of light waves by slits leads to a fascinating phenomenon when a wall has two slits in it that are irradiated with light: In this case, a characteristic pattern is formed called *interference pattern* or *diffraction pattern* (see figure 5).

A double-slit device, or, simply, a *double slit,* is a wall with two parallel and closely neighbored slits in it. When such a setup is irradiated with light, each of the slits becomes a little light source, and the waves spread out everywhere behind the two slits. Because of this, light waves coming through one interfere with the waves coming through the other. At a detector or on a screen set up behind the slits, a pattern of fringes appears (see figures 5 and 6), in which areas of brightness alternate with areas of darkness. In places that are bright, the waves are interfering constructively; in the dark spots, they are wiping each other out. This is the basic message to take from figures 5 and 6: Interference patterns created by a double slit have the form of fringes, lines of brightness alternating with darkness.

Figure 4. THE DIFFRACTION OF WAVES BY A SLIT

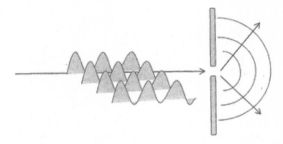

Waves running through a slit in a wall bend around its corners and spread out in all directions behind the slit. The slit becomes the source of circular waves.

Figure 5. INTERFERENCE PATTERN OBSERVED AT A DOUBLE SLIT

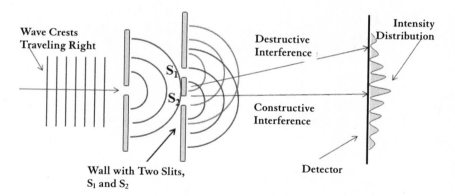

A wall with two slits in it, S_1 und S_2, which are irradiated with light. Because of the diffraction of light, each of the slits is the source of waves that spread out in all directions behind it. Therefore, when two slits are irradiated with light, the waves from one interfere with waves from the other. Along some lines the interference is destructive, the waves cancel, and the corresponding area on the detector is dark. Along other directions the waves reinforce, the interference is constructive, and the corresponding areas on the detector are bright. As a result, an interference pattern is observed on a detector behind a double slit, a pattern of fringes of alternating bright and dark bands.

**Figure 6. INTERFERENCE PATTERN OBTAINED WITH
MONOCHROMATIC LIGHT AT A DOUBLE SLIT**

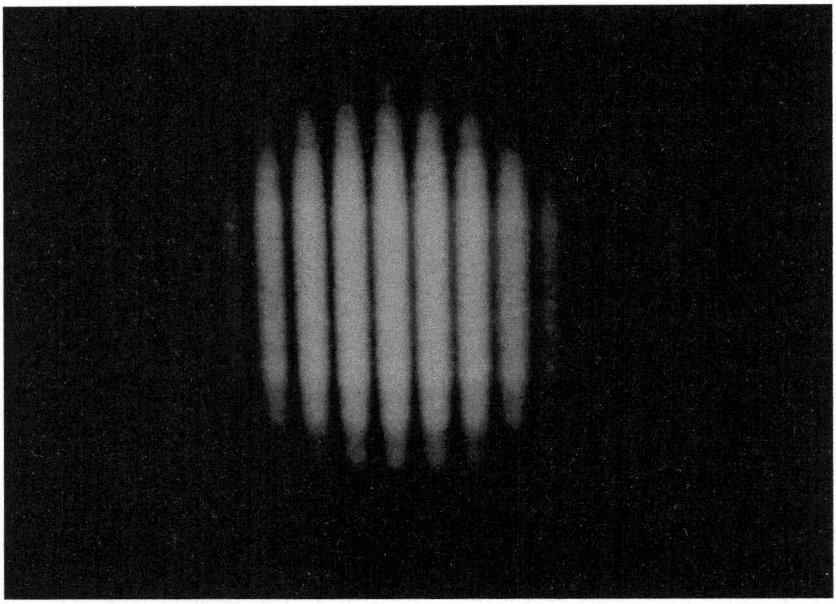

The interference pattern is a pattern of fringes of alternating high and low intensity. In the case of light, high intensity means brightness; low intensity means darkness. (Photo for figure 6 was kindly provided by Wolfgang Rueckner and is reprinted here with his kind permission. For details, see Wolfgang Rueckner and Paul Titcomb, "A Lecture Demonstration of Single Photon Interference," *Am J Phys* 64, 184–88 [Feb. 1996].)

Let's stop for a second and inventory: We have determined that material objects, such as bullets or balls, are typically localized, massy, and impenetrable things. From this it follows that bullets shot through a double slit won't mess around and bend around any corners. Rather, each bullet means business, is on a track of its own, and will either hit the wall or clear a slit and hit the detector, where it will deliver its impact independently of the others. In this process, two piles of bullets are formed behind the slits and after some time the piles add and grow together (see figure 7). Two bullets hitting the detector never cancel. When a bullet pierces your heart, you are dead. Two bullets piercing your heart at the same time won't ever keep you alive: You are double dead!

You can also look at this process by thinking of an hourglass in which the sand grains have two holes to drop through. At the bottom, two small piles of sand grains are formed, and then, after some time, they grow together and form a single pile.

Now we are getting to the crux of the matter. Let's say that we are now making an experiment in which we don't shoot ordinary bullets through a double slit, but elementary particles such as electrons, atoms, or molecules. Apart from the fact that these creatures are tiny things, they have all the attributes of bullets: They are massy, localized, and impenetrable. When they meet in space, they don't create some fancy patterns, but bounce off of one another. They have a well-defined mass and size, and each of them is always observed as a local and individual lump of matter. The mass of an electron, for example, is a ninety-thousand-billion-billion-billionth of a kilogram, where a kilogram is a little more than two pounds. The tiniest thing! Physicists don't know exactly how big electrons are, but it is known that they are smaller than a sphere with a diameter of a million-billionth of a meter. Whenever we observe an electron, it appears as an isolated event, like a tiny dot on a photographic plate or a little flash on a TV screen.

So when electrons are shot through two slits, what do we expect to see? We expect that one electron after another will fly through the slits, that two piles will form at the detector, which after some time will merge into a single pile, as shown in figure 7.

Of course, we can't see electrons directly, because they are so small, but detectors, or so-called multichannel devices, that are sensitive to elementary particles are quite common in laboratories. They typically consist of arrays of a large number of individual channels that show when and where an electron hits.

Figure 7. THE BEHAVIOR OF BULLETS PASSING A DOUBLE SLIT

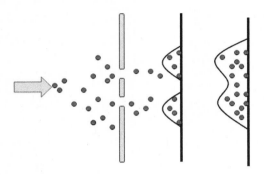

When material particles (bullets) fly through a double slit, they never form interference patterns, just two piles. After some time, the piles will merge and a single pile is formed.

So this is our experiment: Somehow we create a beam of electrons, direct it through a double slit, and watch what happens at a multichannel detector. Such beams can be produced in a vacuum; for example, inside an electron microscope or an electron diffraction instrument that chemists use to study the structures of molecules. Such instruments typically contain a glowing filament (a bent piece of wire like the ones you find in old-fashioned lightbulbs), out of which electrons are dismissed into free space and shot in the direction of targets of interest; for example, toward a double slit. In double-slit experiments using such instruments, the electron current can be adjusted so precisely that at each moment only one single electron is in the apparatus, passes through the slits, and arrives at the detector, without being bothered by other electrons. Only after the detector has recorded a given electron is the next one allowed to enter the game.

When a stream of electrons is running through a double slit, one electron after another strikes the detector and leaves an individual mark, exactly like you would expect (see figure 8).

So at first sight the electrons in such experiments appear to act like normal bullets. However, for some reason it is impossible to aim the flight of electrons at a specific target: These bullets can't be directed, like gunshots, at a particular channel of the detector. Each electron seems to have a certain freedom of choice, where it will show up. If you aim it to the right side of the detector, it may arrive at the left; and vice versa. Each electron acts like it has a stubborn mind of its own and does exactly what it wants to do.

ON SINGLE-PARTICLE INTERFERENCE PATTERNS

When the randomness of the flight of electrons was first discovered, physicists didn't worry too much about it. Their technology had just been developed and, for all practical purposes, the randomness was probably a matter of experimental uncertainties. However, as it turns out, there is a general principle at work here: The behavior of a single quantum particle, such as an electron, is *intrinsically unpredictable* and *principally indeterminate*. Physicists say that single quantum objects are subject to *quantum indeterminacy*. This means that events such as the appearance of a single electron on a detector are completely unpredictable and seemingly ruled by nothing but chance.

So when single electrons are shot through a double slit, each electron appears at the detector in its own way, aimlessly and stubbornly, and decides where it will leave its mark. The single impacts are clearly seen (see figure 8) but, because of the indeterminacy of this process, the sequence of impacts is completely unpredictable. If, for example, an electron has just struck the right side of the detector, the next one may appear in the middle, on the left side, or again on the right.

Figure 8. ELECTRON DIFFRACTION AT A DOUBLE SLIT

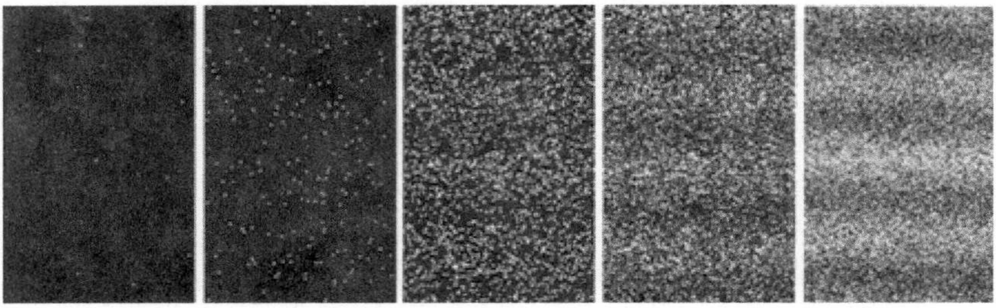

The buildup of an interference pattern in a single-electron interference experiment. In this sequence of pictures, every dot represents the impact of a single electron. Even though the sequence of the impacts is unpredictable and apparently ruled by chance, in the accumulation of many impacts a hidden order appears in an electron interference pattern. (From A. Tonomura, J. Endo, T. Matsuda, T. Kawasaki, and H. Exawa, "Demonstration of Single-Electron Buildup of an Interference Pattern," *Am J Phys* 57: 117 [Feb. 1989]. Reproduced with kind permission.)

At this point the precise choice of words is very important. When we say that the sequence of impacts is unpredictable and ruled by chance, this doesn't mean that it is *arbitrary*. It isn't arbitrary, *because in the accumulation of many random impacts a hidden order is revealed: an interference pattern, as in the interference of light waves at a double slit* (see figure 8). Even though each electron runs through the double slit on its own and arrives at the detector on its own, the single impacts—single random events—are somehow hanging together in a hidden order.

Physicists call this phenomenon *electron diffraction*. Since electron diffraction patterns can be created by the impacts of *single* electrons that passed the slits in isolation, the phenomenon is also called *single-particle interference*.

Single-particle interference can also be described in this way: When elementary particles pass a system of slits, the piles they form on a detector create a contiguous pattern of fringes with high and low intensities. In the case of light (figure 6), high intensity means brightness or a lot of light; low intensity means darkness. In the case of electrons, a region of high intensity means that many electrons have struck there; low intensity means that few or no electrons were observed there.

The Metamorphosis of Matter: Dissolving Material Particles in Nonmaterial Forms

The phenomenon of electron diffraction leads to a number of puzzles. The first puzzle is that the pattern of fringes in an electron diffraction experiment can be explained only by the interference of waves that have come out of the two slits at the same time. Take a look again at figure 5. The up-and-downness of intensities at the detector needs the interaction of something—some sort of wavelike signal that has emerged out of both slits. After all, if one of the slits is closed, the interference pattern vanishes.

But the elementary particles that we call electrons aren't waves or spread-out signals. They aren't objects that can pass a double slit by flying through both slits at the same time. Whenever we see them, they are true elementary particles. Elementary particles can't be broken up into smaller units of matter, and they are localized individuals. Apart from that, at the detector whole electrons are observed in each impact, not some bits and pieces. Moreover, electron particles can't spread out, like glue, over the whole extension of a double slit. Precisely: An electron particle is about ten thousand billion times smaller than the double-slit system through which it is shot. Thus, it is impossible for a single electron particle to split up at the slits and pass through both of them at the same time to create interfering signals. But without interfering signals, there is no interference pattern. How, then, can electrons create a diffraction pattern?

The answer to this enigma is as simple as it is shocking: *When electrons are left alone, they become waves!* Being left alone means that very few objects in its environment exist with which an electron can interact. This is why such experiments are performed in a vacuum, where interactions are relatively rare. We can also describe this behavior in the following way: The single-particle interference phenomena force us to believe that isolated atoms and molecules are subject to dispersion; that is, they flow apart in space.

As it turns out, single-particle interference patterns are not formed only by electrons. All microphysical material objects behave like that. Elementary lumps of matter aren't, like Newton believed, enduring, immutable, hard and eternal things; they spread out spontaneously in fields of waves—they *become* waves—when they are left alone. When the waves spread out in space and pass a double slit, they give rise to interferences, like the interferences that we know are created by ordinary light waves or water waves (see figure 9).

If any material structure, for example some instrument or detector, interacts with the waves into which particles dissolve, the waves will abruptly contract to a point in space that appears as a

---

Figure 9.

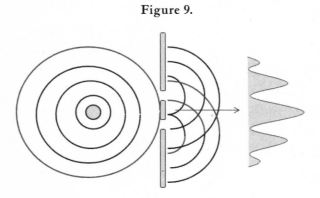

If the wavelike field associated with a particle passes a double slit, this can give rise to interferences that are similar to those produced with ordinary waves, like light waves or water waves.

material particle (see figure 10). This is exactly what happens at the detector level in an electron diffraction instrument: Between the source and the detector, the electrons fill the space of the instrument as waves. At the detector these waves collapse to a point. It is reasonable to think that this behavior is general and universal; that is, we can think that the entire universe is filled with such waves and that all material structures everywhere exist because of the collapse of such waves.

Figure 10.

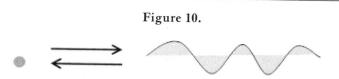

Isolated microphysical material particles dissolve spontaneously in extended fields of waves. In interactions with objects in their environment, these fields contract abruptly to a point, which appears to the observer as a material particle.

So there you have it: Electrons and other material particles such as atoms and molecules can give rise to interference patterns because they can turn into waves when there is nothing in their environment with which they interact, and they do so spontaneously. Since they are doing this when they are on their own, the wave state is, so to speak, their natural state. Wave properties have been observed for many other microphysical objects. The wave properties of neutrons were studied by Roland Gähler and Anton Zeilinger; those of helium atoms by T. Pfau and his coworkers; those of rubidium by S. Dürr et al.; and Markus Arndt et al. were able to study the wave properties of a molecule containing sixty carbon atoms.

At the basis of material things we find invisible fields that actualize in the form of particles whenever we interact with them. Since our interactions always produce material particles, we have been fooled into thinking that matter is eternal. Whenever you look at it, it is a particle. But being a particle is a transitory state of existence, whereas being a wave is the enduring state: It continues as long as it isn't perturbed. Where the matter is coming from, when the wave states of elementary entities become particles, is still a frontier of research. Indications are that the properties of matter emerge out of an invisible field, called the *Higgs field*, which permeates the universe.

This is the metamorphosis of matter: Elementary material particles can make transitions into wavelike states, in which the properties of matter are lost and those of waves appear, as shown by the interference phenomena on double slits. But what is the nature of these waves? We find the answer in the statistical nature of the interference patterns.

THE ORDER OF THE UNIVERSE RESTS ON NUMBERS

We have seen above how single electron impacts in electron interference experiments are unpredictable, in accordance with the general indeterminacy of single quantum events. In contrast to this, the outcome of *many* repeated events can be predicted precisely. You can compare this situation to the life expectancy of a population. For a large population the general life expectancy can be predicted precisely; the exact time of death of an individual person, however, is unpredictable. In electron interference experiments this means that the *intensity distribution* of a large number of impacts is precisely predictable. The intensity distribution determines the general shape of an interference pattern. You can say that, where the intensity is high, the pattern is bright (figure 11; where it is low, the pattern is dark: Along an interference pattern the intensity varies smoothly, going up and down, and up and down. For a single electron, intensity distribution has the meaning of probability—that is, the probability of where it will hit. In areas in which the intensity of impacts is high, many electrons are observed; the probability to be hit by an electron there is high. In areas in which the intensity is low, the chances to be hit by an electron are low. These considerations show that the *interference patterns* in single-particle interference experiments are *probability patterns,* and the interfering waves are probability waves (see figure 11). This interpretation is in agreement with Max Born's discovery that the electrons in atoms and molecules are probability waves. In the late 1920s, Born showed that the squared amplitudes of Schrödinger's wave functions of electrons in atoms and molecules provide probabilities to find the electrons in space.

Now, what are probabilities? Probabilities are dimensionless numbers—ratios of numbers, such as fifty-fifty or seven out of ten. Thus, the single-particle interference phenomena seem to tell us that elementary particles can dissolve in states in which they are dimensionless numbers that spread out in space in the form of waves. By *dimensionless* we mean that the mathematical representations of these waves don't have any physical units—often called *dimensions* by physicists—like units of matter or energy; they are nothing but numbers. Thus, electrons, atoms, and molecules have mass, energy, and a specific position in space only when they are particles. When they become waves, the properties of mass and energy are lost and replaced by the properties of numbers.

Schrödinger's wave mechanics make it possible to calculate the precise *forms* of the waves that produce the intensity distributions observed in interference patterns. When the experimental conditions are changed—for example, when the electrons' speed in which they run through the slits is changed—the *forms of the waves* also change, and the diffraction pattern changes.

At the level of atoms and molecules, a frantic dance is constantly going on: out of matter and into form, out of form and into matter. From such considerations the view evolved that all visible things are emanations out of a field of forms that is spread out through the

Figure 11.

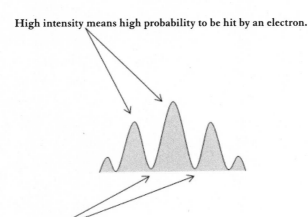

High intensity means high probability to be hit by an electron.

Low intensity means low probability to be hit by an electron.

Understanding the electron diffraction pattern as a probability pattern. Where the intensity is high (in the bright spots) the chances to be hit by an electron are high. Where the intensity is low, the probability to be hit by an electron is low.

entire universe. When material particles dissolve in fields of mathematical forms and patterns of numbers—when they *become* such patterns and forms—they transcend the domain of matter: They become *transmaterial*. From this arose the notion that *the basis of reality is a domain of transmaterial forms, images, or elementary thoughts*. Physical reality is driven by inner images, like a human being is driven by the images in the mind.

Since elementary particles constantly flow apart in fields of waves, why then is there anything stable and lasting at all, and not nothing? The answer is this: There is a world of stable and lasting things because elementary particles enter the realm of forms only as long as they are free of interactions with other objects in their environment. *It is interactions that actualize; it is interrelations that make reality visible.*

PROBABILITY WAVES FORM AN INVISIBLE REALM
OF POTENTIALITY

When a material particle flows apart in a wavelike state, it effectively leaves a localized existence in space, and we have to wonder *where* it is.

We have a little thing here, a tiny lump of matter. We have just seen it at a certain point in space: It is real. But all of a sudden it flows apart and becomes a wave. We don't see the wave, but we know that it must be there. So it is a meaningful question to ask: Which position in space does an elementary particle occupy when it is a wave? The answer is simple but amazing.

Like all waves, probability waves are extended in space. Thus, when a material particle becomes such a wave, it has no specific position in space: *It is, actually, nowhere.* It occupies a specific location only when it is a visible, material particle. In its wave state, its position in space has no *actual* location, but many *potential* locations. With this, the magic word has entered the scene: The wave states into which microphysical objects dissolve are *potentiality states.* Since these are the natural states for microphysical objects to exist in—the states into which they will make spontaneous transitions, whenever they are on their own—we can conclude that the visible reality emanates out of a realm of potentiality that is underlying all things.

This is how we are led to the view that physical reality appears to us in two domains: the realm of the *actuality* of localized material things, and the realm of *potentiality* of nonmaterial forms that are spread out in space. These forms are real, even though they are invisible, because *they have the potential* to manifest themselves in the empirical world and act in it.

The notion of potentiality is naturally connected with the notion of probability, because the probability of presence tells you where a particle *can* be found. Indeed, if a number of detectors are

set up in different locations in space to catch an electron in its potentiality state, all of a sudden it will pop out in one of them where
the probability to find it is non-zero. In this way probability waves
are also potentiality waves.

On the Concept of Superposition States

Figure 5 shows that the interference pattern in double-slit experiments is the result of the *superposition* of waves that came through
both slits at the same time. This observation can lead you in a simple way to the concept of *superposition states.*

Consider for a second that an electron in a double-slit experiment behaves like an ordinary gunshot bullet that follows a specific
trajectory through the two slits, which means that it passes through
just one of them. We can call this trajectory a *state* of the electron. If
its state is the trajectory through slit no. 1, we will call this the "state
no. 1," or, symbolically, S_1. Similarly, if the trajectory of the electron
is specifically through slit no. 2, we call this "state no. 2," or S_2. That
the interference pattern is formed by signals that passed through
both slits at the same time must mean that a single electron must
have traveled in a *superposition state,* which is the sum $S_1 + S_2$. The
meaning of such a state isn't that the electron travels through both
slits at the same time, or takes all possible trajectories, but that it
has taken no actual path: *Its path is indeterminate.*

If you find this statement questionable and want to check it out,
you can install two detectors, one in each slit, to find out through
which slit an electron is traveling. In that case, an electron will always be found in just one of the slits—never in both at the same
time. However, when it is known which slit it has come through,
the interference pattern breaks down.

In this context physicists often speak of *complementarity*: path
determination or position determination and interference exclude

each other. You can know which way an electron travels, but then there is no interference pattern. Or you can observe an interference pattern, but then you can't know which way the electron traveled. Whenever the system contains information on the path or the state of a particle, the interference vanishes, even when the information is collected in a way that doesn't disturb the system in any way. In this context we also say that, when it can contribute to interference phenomena, the electron is in a coherent state, where *coherence* is the ability to interfere. Which-way information and coherence exclude each other.

Superposition states, like the superposition of many paths, are actually states of potentiality: When you interact with such a state, one of the states contained in the superposition will be actualized, but you can't predict which one. This is often referred to as the *collapse of the wave function* of a superposition state. The expression describes that the many states contained in a superposition collapse into a single one. The inability to predict exactly how a superposition will collapse in a measurement is often called the *measurement problem* of quantum physics.

It is sometimes said that an electron in a double-slit experiment, in which it is in a superposition state, $S_1 + S_2$, takes both paths, but that isn't correct. In a superposition state, a particle is actually taking none of the paths, because it isn't in any actual state, but in a state of potentiality in which it isn't a part of the empirical world.

The inability to predict with certainty which one of the states contained in a superposition will actually appear in the empirical world is at the core of *quantum indeterminacy.* Quantum indeterminacy can also be described as the inability to predict individual quantum events.

At first sight it is possible to think that the unpredictability isn't such a big deal. You know many events in everyday life that are unpredictable, and it doesn't matter! You can't predict the outcome, for example, when you throw a die or toss a coin. You know that the chances for heads or tails are fifty-fifty in many tosses, but you can't predict the outcome of a single toss.

Here it is important to realize that the indeterminacy of quantum events, such as the movement of a single electron, isn't like the unpredictability of tossing a coin. The tossing of a coin has its causes, like the way in which you move your hand, but the outcome is unpredictable, because *our information is incomplete* as to how the coin was tossed. If we knew exactly the momentum transferred by your hand when it tossed the coin, and what the density of the air through which the coin had to fly was, and so on, the outcome of each single toss could be predicted with certainty. The path of a tossed coin has its causes, but we don't know them because there are too many of them. In contrast, the behavior of a single electron, or of a single state in a superposition, is unpredictable not because our information is incomplete, but because *it has no causes; there is nothing there to know*. If causes exist, they belong to the nonempirical part of reality, and we can't see them! In this way the quantum phenomena open the door to influences from a transcendent part of reality.

THE CONCEPT OF NONLOCALITY

When you watch the news on television, it isn't unusual to see a reporter at the other end of the world describe some event that just happened there. When the anchor in Washington asks a question, at first the distant reporter just stands there, doing nothing and rolling his eyes, until, all of a sudden, he is jerked into action when the question arrives in his earphone. This is an example of how, in the world of ordinary things, no signal is able to travel at a speed faster than the speed of light. To start some action somewhere else far away, you have to wait at least as long as it takes for a signal to get there. There is always an unavoidable delay between question and answer when you are talking to an astronaut far out in space. In this regard, too, the quantum world is different: Under certain conditions influences can act instantaneously over arbitrarily long distances, as though the time between the two events or the space between them didn't exist. Also in this regard, the particle interference phenomena can serve as an instructive example.

Think about the state of an electron in a single-particle interference experiment immediately prior to its impact on the detector. The multichannel detectors used in such experiments typically have thousands of independent channels, and the electron can appear in any of them. This means that, immediately prior to being observed, the electron is in a state in which it has the potential to appear in any one of the detector channels; each channel has a chance to catch it. At the exact moment when it hits in one of the channels, the probability to find it there is equal to 1, or a 100 percent, and instantly zero in all the other channels. This is an example of a *nonlocal* event: Something you do here and now—namely, catching an electron—has an instantaneous effect somewhere else, an arbitrarily long distance away—namely, changing a local potentiality.

A free electron can be used as another example. A free electron is one that can move through space without being subject to potential energy. For example, when an electron leaves an atom, it enters into such a state and is free to move anywhere it wants. Such particles can evolve in states in which their probability of presence is spread out through large areas of space. In such a state, the position of an electron doesn't have an actual value, but many potential values. The electron is, actually, nowhere.

No empirical object can exist in such a state. Think of one of your friends who is not with you right now, and ask yourself where she is. Even though you may not know where she is, you know that she is somewhere. Empirical things are always somewhere. If an object is at, say, point A in the universe, the probability to find it there is equal to 1 (unity) or 100 percent. This is *essentially* different from a probability of presence that is 10 percent at point A, 5 percent at point B, and so forth.

If we decide, at this point, to take the strange behavior of this electron personally and go after it, we can set up several detectors in the laboratory, trying to determine where it is. We turn the detectors on and wait. All of a sudden it will appear in one of them, but we can't predict in which one. As soon as it appears somewhere—say, at point A—the probability at that point will be 1 or 100 percent, and *instantly* the potentiality will be zero everywhere else in the lab.

All transempirical states are like that: They can fill large areas of space in potentiality and when something is done to them here and now, instantly the state will change everywhere else.

Appendix for Chapter 2

HOW THE NONEMPIRICAL PART OF REALITY IS DISCOVERED IN THE VIRTUAL STATES OF ATOMS AND MOLECULES

In this appendix you will find definitions and descriptions of some basic concepts of quantum physics and quantum chemistry. The term *quantum* will be explained, as it is used in *quantum physics, quantum states,* and *quantum jumps.* Among the quantum states of atoms and molecules the meaning of the *occupied states* and *empty states* will be considered in detail. The empty states are often called the *virtual states* of atoms and molecules, and their properties will be described using some simple examples.

QUANTUM MONEY

In Latin the word *quantum* (plural *quanta*) means something like "so much." In physics, a quantum is the smallest indivisible amount or quantity of some property, such as an electric charge or energy. The electric charge of an electron, for example, is an indivisible elementary unit. When you pay your utility bill at the end of the month, you pay, in a way, for the number of electric charges that you took out of the power line. Electric charge comes in the form of smallest indivisible units, or elementary charge quanta. The charge on an electron is such a quantum, and you can't divide it into smaller parts. You can't ask your power company to sell you a third of the electric charge of an electron.

The same happens with energy. In ordinary light, like sunlight, energy depends on the frequency of the light waves. Visible light, for example, has less energy than ultraviolet light, because visible light frequency is lower than ultraviolet light frequency. On the other hand, visible light carries more energy than infrared light, because its frequency is higher than that of infrared light. Frequency in visible light also determines the color of light. Red light has a lower frequency than blue light. High-frequency ultraviolet light can destroy the cells in your skin and cause cancer. Infrared light doesn't do that; it only warms your skin.

Money, too, comes in quanta. If you have a bundle of dollar bills, you can pay out five dollars, or five hundred, but you can't pay five dollars and ten cents, because dollar bills are quantized. You couldn't pay ten cents by cutting a piece off a dollar bill.

The example of money also shows that quanta exist in various sizes. Instead of dealing with a bundle of dollar bills, you can use a box of dimes, for example. Now you can pay out smaller quanta, or smaller amounts of money that are less than a dollar, but, in themselves, dimes are also quantized: They are quanta of ten cents. You can't use them to pay thirty-three cents, or seventy-nine.

As it turns out, exactly the same condition applies to light. The energy that can be taken out of a beam of light comes out in the form of light energy quanta, called *photons*. The quanta can have different sizes, depending on the frequency of the light waves, but for each frequency there is a smallest, indivisible amount of energy that you can take out of a light beam. Physicists call any beam of light whose waves all have the same frequency or color *monochromatic*. The energy that you can take out of such a light beam doesn't flow out of it, like a liquid, but jumps out in discrete units—that is, in the form of quanta. The quanta coming out of a monochromatic beam of light are all exactly the same size. You can take out one quantum or two, but not one and a half or one-tenth. Quanta are indivisible. It is as though a beam of light was a stream of coins. If it has a frequency that corresponds to a stream of nickels, you can take out five cents or twenty-five, but not thirty-three or twenty-nine. If it has the frequency corresponding to dimes, you can take out a whole number of dimes. High-frequency light has bigger quanta than low-frequency light. But whatever the quanta are of a given frequency, they are always identical and indivisible.

The energy quanta of light—the photons—represent the particle state of light. Like electrons and atoms, light also has a wave state and a particle state. In a beautiful study, Wolfgang Rueckner and Paul Titcomb have shown that the arrival of photons can actually be seen in dim light using a video camera of extreme light sensitivity. In interference phenomena, light displays the properties of waves. When it exchanges energy, it gives off lumps of energy—quanta—like little balls, and it displays the properties of particles. These particles can push and pull and bounce around, like all other particles do.

On Energy Ladders and Empty Steps

Light that appears as white light in the human perception is a mixture of many types of light waves with different frequencies. In the language of the last paragraph, white light is a stream of photons of all kinds, or of energy quanta of many sizes.

Irradiated with a beam of white light, atoms or molecules can take energy out of such a stream of photons, but they are very picky as to what kind of quanta they will fish out of a streaming mixture. A molecule is like a Coke machine that will accept quarters and dimes but not nickels and pennies. A molecular *spectrum* is a representation of all the frequencies that a molecule can absorb and of the relative intensities of the absorptions. *Spectroscopy* is the field of study in physics and chemistry that deals with atomic or molecular spectra. Each spectrum is unique—like a fingerprint—and can be used to identify the molecule that produced it. The selectivity of absorption means that the energy that a system can exchange with its environment is quantized.

You can understand that, inside an atom or molecule, the sudden arrival of a lump of energy must cause sudden and disruptive changes. Since it can't be swallowed in little sips, like a cup of tea, but only as an indivisible quantum, the arrival of a photon inside an atom or molecule is like the impact of a bullet; it causes a jump in the inner state of a system. Thus, their way of interacting with light is indicative of a structure inside the atoms and molecules that makes jumps possible: It confirms Bohr's insight that all things exist in quantum states—that is, in states with a fixed energy, and when they absorb a photon, they use its energy to jump from a lower state to a higher state.

The quantum states in atoms are energy states. They form a ladder of energies. We can symbolically represent a quantum state by E_n, where E is for energy, and n is a number, called a *quantum number*. The lowest state or the ground step of the ladder is E_1; the next higher state, E_2; and so on. An example of such an energy ladder, E_1. E_2. E_3 . . . E_n, is shown in figure 12. Seven levels are shown in figure 12, from n = 1 to n = 7. In real atoms, the quantum number n can assume any whole-number value, so that each atom and molecule contains countless numbers of quantum states.

When you are using a ladder to paint a wall in your home, you have to stand on one of its steps; you can't float in the air between two steps. Exactly the same condition applies to atoms and molecules. A hydrogen atom, for example, has a single electron, and under normal conditions that electron must stand exactly on one of the steps of the electronic energy ladder of that atom; it can't hang suspended between several steps. We say that the electron in the hydrogen atom must occupy one of the quantum states of that system, and it can't float in between two states. *It follows that all but one of the countless quantum states of a hydrogen atom are empty.*

For atoms and molecules with more than one electron, the situation is more complicated, but basically it is the same: Each electron must occupy exactly one of the energy steps or quantum states of a given system. You can call the combination of states that are occupied in an atom or molecule with many electrons a *state* of this system. Always one state of a thing is occupied while countless others are empty. When we observe a system, we always see it in an occupied state. The manifold of the empty states is also there, but invisible. Quantum chemists call empty states *virtual*.

The Hydrogen Atom as an Example

Let's take a closer look at the hydrogen atom.

A hydrogen atom is a simple system. It consists of two elementary particles, a proton and an electron. The proton is the nucleus of the atom. It contains all of the positive electric charge and nearly all of the mass. It is about 1,800 times heavier than an electron. The electron is somehow outside of the nucleus: The question is, how?

In 1926, Austrian physicist Erwin Schrödinger, inspired by the wavelike behavior of elementary particles (see our descriptions of the ETs in chapter 1), developed a version of quantum theory in which a physical system is represented by a *wave function*. To describe the mechanics of atoms and molecules, Schrödinger set up a wave equation that is similar to the equation used in optics to describe the propagation of light waves. Physicists find it convenient to use symbols in their discussions; the Greek letter Ψ (pronunciation: "psi") is usually used as a symbol for Schrödinger's wave functions.

When Schrödinger's equation is solved for an electron in a hydrogen atom, a whole spectrum of solutions is obtained, consisting of countless wave functions, Ψ, each of which defines a quantum state in which the electron in the hydrogen atom can exist. Each Ψ represents a mathematical expression that defines a specific waveform that determines the physical properties of the state that

Figure 12.

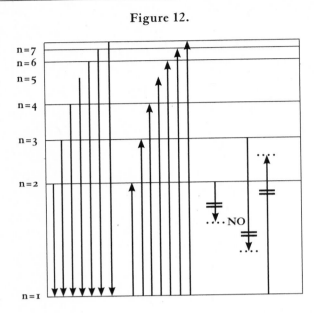

Energy level diagram, or energy ladder, of the type found for the quantum states of atoms. In this figure the height of each step in the energy ladder is determined by a quantum number, n, and energy levels from n=1 to n=7 are shown. An atom can make state transitions, jumping from one level to another, as indicated by the arrows. The corresponding energy differences are exchanged in the form of photons. If the transition is from a lower to a higher state, a photon with exactly the right energy must be absorbed. If it is from a higher to a lower state, a photon is emitted. Transitions to levels between two steps are forbidden; for example, forbidden transitions are shown from n=1 to somewhere between n=2 and n=3, or from n=2 to somewhere above n=1.

it represents. The waves themselves are invisible, but their squared amplitudes, $|\Psi|^2$, correspond to observable properties of the electron: They determine, for each state, the probability of presence—that is, the probability of finding an electron in the space surrounding the nucleus. In this context physicists often speak of *probability density,* meaning the probability to find an electron in a unit volume of space.

The squared wave functions describing the state of electrons in atoms or molecules are often referred to as *orbitals*. The term derives from the time when physicists discussed Bohr's model of fixed electronic orbits, in which the electrons were thought to run around nuclei, like the planets revolve around the sun. But such a motion is impossible. There are strong attractive forces between a negatively charged electron and a positively charged atomic nucleus. Because of these forces, an electron can't revolve around a nucleus in a stable orbit, but must crash into it and stay put there. So, since it isn't in orbit, but somehow outside the nucleus, the term used to describe the states of electrons in atoms and molecules was changed from orbit to *something like an orbit,* or an *orbital.*

It is worthwhile to stop for a second and consider the picture that Schrödinger's theory is presenting of the nature of matter. The theory implies that the electrons outside the atomic nuclei are waves. These waves are "standing" waves in the sense that they don't propagate in space, like the water waves in an ocean; they are tethered at the nuclei. Since the meaning of these waves is that of probability waves, they are numbers. Thus, the electrons in atoms are numbers, mathematical forms, or numerical patterns. The wave theory doesn't mean that an electron in an atom is running around the nucleus in waving circles, like in an epicyclic motion of Ptolemaic planets. Rather, for all practical purposes, when an electron enters an atom, it becomes a probability wave; that is, it has ceased to be a material particle and has stepped into a different state of being, like the wave states of ETs whose actions we saw in appendix 1 in the single-particle interference phenomena.

It is important to understand that, when an electron enters an atom and becomes a nonmaterial wave, that doesn't mean it has been annihilated. It means only that an ET has undergone a transition from a particle state to a wave state. The wave state is as real as the particle state, even though it isn't a material state. From the wave state the particle can pop out again. That is, transitions back and forth are always possible, so that the electron can constantly dance in and out of the empirical world.

As a probability wave, an electron in an atom has no specific position in space, but many potential positions. It is in a state of potentiality in which it has the potential to appear in many places but is actually in no specific place. This is as we defined the concept of potentiality in appendix 1.

THE CONCEPT OF VIRTUAL STATES

At this point we can further deepen our understanding of the potentiality in atoms by considering the shapes of some orbitals. David Manthey has created a wonderful website (http://www.orbitals .com) that allows you to calculate and plot any desired orbital of the hydrogen atom. His program was used to generate the pictures of the orbitals shown in figure 13.

To begin with, let's talk money. Let's say you are buying some groceries, and the amount that you have to pay is $2.25. In looking at this number, you could say that we need two types of numbers to specify an amount of money. In $2.25, the first number, 2, defines the quantity of dollars that you have to pay. We can call this a quantum number for dollars. The second number, 25, defines the quantity of cents that you have to pay. It is a quantum number for cents. In this way, amounts of money depend on two quantum numbers.

I don't expect that the introduction of quantum numbers into financial affairs will soon be rewarded with the Nobel Prize in Economics. Nevertheless, for our present purposes the example is instructive, because something like this applies to the amounts of energy associated with quantum states: They depend on quantum numbers. For the hydrogen atom three numbers are needed to define a state, and this is the meaning of the triplets of numbers, 1.0.0,

2.1.0, 3.2.0, and so forth, shown in figure 13. Each triplet identifies a specific quantum state, its energy, and the shape of its orbital, $|\Psi|^2$.

A hydrogen atom contains countless quantum states whose properties are determined by the waveforms of their orbitals, which, in turn, are determined by the physical conditions of the system. Some typical orbitals are shown in figure 13.

When you look at a hydrogen atom, you will always find it in one of its states. Let's say you have caught such a thing and as you are holding it in your hands, you find it in its lowest energy state, the 1.0.0 state. When it is in this state, the higher energy states also exist—that is, they are present in this atom, but they aren't visible because they are empty: There is nothing there to see. But the logical order of the empty states exists because it is a part of the logical constitution of an atom, and it contains the future empirical possibilities of an atom. All that the atom can do in the future is actualize one of its empty states. When an atom jumps into an empty state, that state becomes visible. Because it *can* manifest itself in the empirical world, the empty order belongs to the realm of potentiality in the physical reality.

Quantum chemists call empty states *virtual*. Virtual states are real, but since they are empty, they are nonempirical. You can think of them as mathematical forms, wave functions, or probability patterns, but they are more than graphical images or photos of forms or mathematical formulas. They are truly existing potentiality. They can actualize in the world of material things and act in it: When an atom jumps into a virtual state, that state ceases to be virtual and becomes visible.

When we observe an atom, its visible properties are those of the state that it occupies when it is observed. A hydrogen atom has different physical properties in the 1.0.0 state than in 4.3.0 state or any other state. So when it jumps from 1.0.0 into 4.3.0, the properties connected with a spherical orbital will vanish, and those of a doughnut-like orbital will appear (see figure 13). Like the hydrogen atom, all atoms and molecules have the potential to actualize their virtual states, and in this process new forms can appear in the empirical world. It is in this way that the virtual states contain the future empirical possibilities of the world. Nothing can appear in the world that doesn't already exist virtually somewhere. New and complex forms can't appear out of nothing, because quantum systems can't jump into nothing.

You might wonder at this point why we have to say that virtual states are invisible. After all, their energy is known, and the forms of the orbitals are known. Why do we have to think that there is nothing there to see? Being empty or full doesn't affect the visibility of other things, such as glasses or bottles, for example. So why should it affect the state in which orbitals exist?

The answer lies in the physical meaning of the orbitals. Remember that the orbital of a state provides the probability of pres-

---

Figure 13.

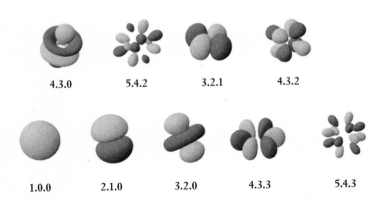

| 4.3.0 | 5.4.2 | 3.2.1 | 4.3.2 |

| 1.0.0 | 2.1.0 | 3.2.0 | 4.3.3 | 5.4.3 |

Atomic orbitals (iso-density surfaces of probability distributions, $|\Psi\, n,l,m|^2$) of some selected atomic $n.l.m$-states. (The Orbital Viewer program by David Manthey, http://www.orbitals.com, was used to calculate the orbitals, with kind permission of David Manthey.)

ence for an electron in this state. If a state is empty, it means that it doesn't contain any electrons. Thus, there is nothing there for which a probability of presence could be observed.

THE EXAMPLE OF A PARTICLE IN A BOX

Physicists call a mass particle whose motion is constrained to the inside of a box, or a deep well of potential energy, a *particle in a box*. You can think of a ball that has fallen into a canyon or a deep ravine. A particle in a box can serve as another simple illustration of the properties of virtual states.

When a particle is in a potential energy hole, it has to exist in quantum states in which its energy is quantized. As in the case of the hydrogen atom, the quantum states of a particle in a box form a ladder of energy, and each step is characterized by its wave function, Ψ, and probability density, $|\Psi|^2$. When the particle is confined to move in two directions, such as in different directions on the floor of the box, two quantum numbers are needed to define a

Figure 14.

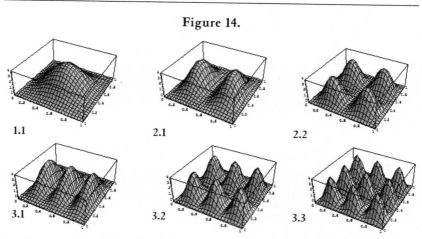

Iso-density surfaces of probability distributions, $|\Psi|^2$, for the quantum states of a particle in a box.

state. Figure 14 shows how the $|\Psi|^2$ of each state fills the box with the typical maxima and minima of waves. Like electrons when they enter a vacuum or an atom, an unobserved particle confined to a box is a wave.

Figure 15 shows the specific case of the probability density of the 10.10 state of a particle in a box. It reveals a characteristic feature of quantum states that is significant and very instructive.

As is seen from figure 15, the 10.10 state of a particle in a box displays a large number of probability maxima—areas where it is likely to find the particle—that alternate with areas where the probability to find the particle is zero. Such a probability distribution creates a peculiar situation. Each maximum is isolated from the next by a boundary of zero probability. The particle must *never* be in a zero probability region; it is an absolutely forbidden region! Nevertheless, in repeated observations, you can be sure that the particle will sometimes be found in one of the maxima, and then in another. But to go from one to the other, it has to pass a region of space where it must never be! How does it do that?

To appear at different points in a box, an ET must blink in and out of the empirical world. When we enter a room and close the door, we are practically a particle in a box. Does this mean that when we walk from one end of the room to another, we are blinking in and out of the empirical world and we don't even know it?

This situation is typical for quantum systems. Probability densities are often structured into areas of space that are separated by planes of zero probability—so-called *nodal planes,* where the particle must never be, not even for a fleeting moment of passage. The 2.1.0 state of the hydrogen atom, for example, consists of two disconnected spheres that are separated by such a nodal plane (figure 13). Nevertheless, in repeated observations of the 2.1.0 electron, it will sometimes appear above that plane, and sometimes below it. It is a mystery how a visible particle can get from one area in space to another, necessarily passing and not passing a forbidden space!

Figure 15.

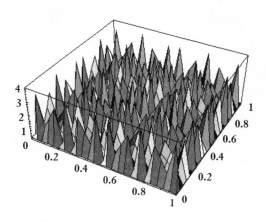

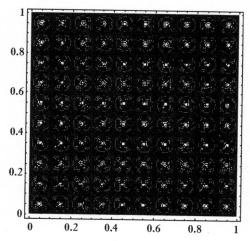

Iso-density surfaces of probability distributions, $|\Psi|^2$, for the 10.10-state of a particle in a box. A three-dimensional rendering is shown in the figure at the top; a contour plot is shown at the bottom. Bright areas in the contour plot are probability maxima. Dark spots are regions of zero probability.

How magical this situation is can be illustrated by a simple thought experiment. Let's assume that some tyrant has constructed an infinitely large and impenetrable steel wall that divides the universe into two disconnected countries: the Right-Side Republic and the Left-Side Republic. Let's further assume that we have been put in charge of finding a precious gold ball that has somehow been lost but is known to be somewhere. Both countries are eager to find it.

Two teams set out to look for this thing: one in the Right-Side Republic and one in the Left-Side Republic. Once this ball is found on one of the sides of the impenetrable cosmic wall, it will have to stay in that republic forever; it can't all of a sudden pop out on the other side. But this is what electrons and all elementary particles do all the time. Nodal planes in probability densities are like infinitely large and impenetrable steel walls. Real particles couldn't ever diffuse through such walls, but elementary particles do it: At one time an electron pops out on one side; and next time, on the other. How is that possible? It is possible because electrons aren't ordinary material things.

MOLECULAR VIRTUAL STATES

Like atoms, like molecules.

In molecular quantum theory, virtual states are also a well-defined concept. Molecular virtual states arise when atoms form a chemical bond. In a chemical bond atoms let the wave functions of their electronic states interfere with one another. In the interference the wave functions of atoms turn into the wave functions of a molecule—we say that atomic orbitals turn into molecular orbitals. As in the case of atoms, each orbital is defined by its waveform and the fixed amount of energy that must be paid to place an electron into it. When two atoms form a chemical bond, they create a system of molecular orbitals. Like their atomic cousins, the molecular orbitals also form a ladder of energy states, each of which can accept a maximum of two electrons. When it contains two electrons, we say that an orbital is *filled*.

In the quantum mechanical model of chemical bonding it is envisioned that the electrons from each atom in a molecule will be placed, one by one, into the molecular orbitals, first filling the state of lowest energy and then proceeding to the next higher levels as needed. This process leaves a large number of states empty, like the empty rooms in a sparsely booked hotel, since all molecules have many more states than they need to place their electrons. As before, the empty states of molecules are virtual states that aren't part of the empirical world, but have the potential to appear in it. They exist in some kind of purgatory, "between the idea of an event and a real event," as Werner Heisenberg described it.

The hydrogen molecule can be used as a specific example. It contains two hydrogen atoms (chemical symbol H), which are bonded together. Accordingly, the chemical formula for the molecule is H_2.

Chemists call the 1.0.0 state of the hydrogen atom the *H1s state*. Let's use this notation for reasons of brevity. If the wave functions of the H1s states of two hydrogen atoms interfere with each other, they form two molecular states. One, called 1σ, has an energy level below the energy level of H1s, while the other one, called $1\sigma^*$, has an energy level higher than the level of H1s. Each hydrogen atom contributes a single electron to the molecule. In the most stable state of H_2, the two electrons occupy the 1σ state. This leaves the upper state, $1\sigma^*$, empty or virtual. An energy level diagram of H_2 is shown in figure 16.

When an electron jumps into the virtual state of a molecule, that state becomes actualized, and its virtual order becomes a visible order. In this simple way transcendent order can express itself with ease in the visible world. All molecules are centers of potentiality, and constantly new visible forms are evolving out of their virtual states.

Figure 16.

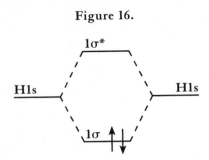

Energy level diagram of the two lowest energy states of the hydrogen molecule, H_2. If the wave functions of the 1s-states of two hydrogen atoms (H1s) interfere with each other, two molecular states are formed. The energy of one of them, 1σ, is below the atomic energy level, H1s; the second molecular energy level, $1\sigma^*$, is higher. When the two electrons of H_2 occupy the lowest state, (1σ), the higher level is empty and virtual. In this figure the two electrons in 1σ are symbolized by two arrows pointing in opposite directions. This presentation is customary to indicate that electrons have an intrinsic spinning movement, which aligns in opposite directions, when two of them occupy the same quantum state.

EMPTY VALLEYS IN INVISIBLE MOUNTAIN RANGES

When atoms interact and form a molecule, the interaction doesn't create only molecular electronic states, but it also leads to a new type of motion: the oscillation of the atoms in the molecule about their most stable positions, the so-called *equilibrium positions.*

You can think that the electronic quantum states in a molecule are energy holes that are formed when the wave functions of two or more atoms interfere with each other. When the atoms drop to the bottom of such a hole, they are bonded to each other, because they have lost the energy that they need to climb back up again, over the hill and into freedom. In the following description I will use the case of a diatomic molecule as a simple example.

Quantum chemists call molecular states *electronic states* if they are formed by the interactions of the electronic orbitals of atoms.

Each molecule has a huge number of such states: A few are occupied by electrons; the majority are empty. You can think of the state landscape of a molecule like a mountain range—perhaps the Rocky Mountains or the Himalayas—with countless valleys and different peaks and elevations. The only thing to remember is that the mountain ranges in molecules are invisible. But that doesn't mean they aren't a part of the landscape. We call the lowest electronic state of a molecule its *ground state*. The higher lying states are called *excited states*.

At the bottom of a valley, the atoms of a molecule don't simply roll to a point of rest and lie there motionless. Rather, they swing about each other in a never-ending motion, back and forth and back and forth, like an eternal pendulum: Physicists say that the atoms in molecules oscillate or vibrate.

All vibrational motion is connected with an energy that is quantized. In each molecule, the vibrational energy defines a ladder of energy whose steps depend on a quantum number. The vibrational quantum number is usually denoted by the symbol *v*. Its allowed values are the whole numbers v = 0, 1, 2, 3, and so on. Each step of the ladder represents a vibrational state. As before, the properties of each state are determined by a wave function. The vibrational wave functions of a diatomic molecule are simple wavelike (so-called sinusoidal) forms.

At this point it will be helpful to take a look at some pictures. Figure 17 shows two hypothetical electronic states of a diatomic molecule and their vibrational energy ladders. The typical outlines of the electronic energy holes are shown in figure 17 for the electronic ground state, E_0, and a higher lying electronic state, E_1. Some of the steps of the vibrational energy ladders in E_0 and E_1 are also shown, depending on the vibrational quantum numbers, for which we will use the symbols v and v', respectively, in E_0 and E_1.

Figure 17 can be used to illustrate what is meant by the statement that *molecules exist in quantum states*. In our example, this

Figure 17.

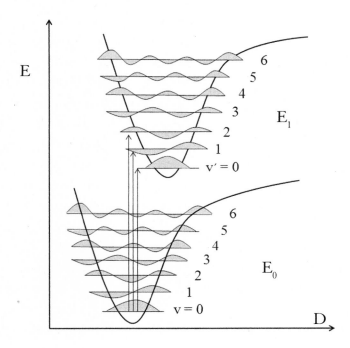

Graphical presentation of the typical outlines of the energy valleys of two electronic states in a diatomic molecule. The lowest state, E_o, is the electronic ground state of the molecule. In the energy field of the molecule, E_1 is a higher lying, excited electronic state of the molecule. Vibrational energy ladders are shown, depending on the vibrational quantum numbers, v, in E_o; and v', in E_1. Energy levels and their wave functions are shown for v=0 to v=6; and for v'=0 to v'=6. The outlines of the energy holes are a function of the distance between the two atoms. When the atoms are very far apart, the energy is high. As they approach each other, the energy decreases until it reaches a minimum. If the two atoms at this point are still pressed closer together, repulsive forces will lead to a rise in energy. The horizontal axis in this figure is a measure of the distance, D, between the two atoms. The vertical axis is for the energy, E, of the molecule. The arrows indicate a possible transition from the ground state, v=0, to higher lying states in E_1.

statement expresses the rule that each molecule must occupy a step on a vibrational ladder in one of its electronic states. This rule implies that the only thing a molecule can do is jump from a state that it occupies into another that is empty. A system is able to change only if it has empty states into which it can jump. Without empty states it would be frozen into itself.

Figure 17 also illustrates what kind of strange things the empty states in molecules are: They are contained in a molecule as fixed structures—but as mathematical, not visible, structures. They are like empty valleys in an invisible mountain range. To see how strange that is, just think that the Rocky Mountains would consist of a single visible valley, with all the others invisible. This is exactly what molecular states are about: When they aren't occupied, they are empty and there is nothing there to see. When a molecule occupies its ground state—$v = 0$ in E_o in figure 17—then the higher lying states also exist, because for each of them the energy and the form of the wave function are already anchored in the molecule before they are occupied.

From a state that it occupies, a molecule can make a spontaneous jump into any empty state that is part of its system. The only condition is that the energy balance must be true. If a molecule is, for example, in its ground state—in $v = 0$ in E_o, in figure 17—then it can make transitions into any higher lying state—for example, $v' = 0$, $v' = 1$, etc., in E_1—provided it finds the energy needed to pay for such a transition. When it is irradiated with light, this energy can be provided by photons whose energy fits the bill. At this point we must remember that the energy of photons depends on their frequency. Out of a beam of light, a molecule can absorb all those kinds of photons whose energies correspond precisely to the energy differences between two of its quantum states.

In this context, the concept of the *transition probability* is of great significance. It denotes the fact that the state transitions of a molecule that are possible aren't all equally probable, even if there is an abundance of energy available for all of them. Independent of energy considerations, some state transitions seem to be easier for a molecule to actualize than others. As a consequence, each transition is characterized by the probability with which it can be observed. This transition probability can be precisely calculated using quantum theoretical techniques.

As it turns out, the transition probabilities between quantum states depend on the wave functions of the states involved in a transition, including the wave functions of the empty states. Differences in transition probabilities mean that, out of a mixture of light with different photons, some photons are absorbed more frequently than others. Using special types of instruments, called *spectrometers,* physicists can determine how many out of a given number of identical molecules will make transitions between their various states. Such numbers define relative transition probabilities. In spectroscopic studies, transition probability is related to measured *absorption intensity.* When the intensity of an absorption is high, the probability of the corresponding transition is high.

So in a nutshell: The relative frequency with which photons are absorbed by a molecule is related to transition probabilities. In a spectrometer, probability is related to intensity. In the absorption spectrum of a molecule, frequently absorbed photons appear with a high intensity. In the hypothetical absorption spectrum shown in figure 18, the transition from $v = 0$ to $v' = 3$, for example, is very likely, while the transition from $v = 0$ to $v' = 4$ practically doesn't occur.

With this we have reached a crucial point of our analysis: Transition probabilities between molecular quantum states—that is, observed spectroscopic absorption intensities—depend on the wave functions of virtual states. In physical chemistry, this principle is known as the Franck-Condon principle. Prior to actually making a jump into an empty state, a molecule explores, so to speak, the aptitude of that virtual state, or its fitness to serve as a target of a

Figure 18.

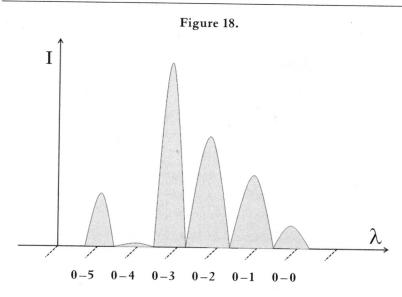

Typical intensity distribution of a molecular absorption spectrum. The figure shows the hypothetical transition intensities of a molecule for transitions between the states $v = 0$ and $v' = 0, 1, 2, 3, 4,$ and 5 (see figure 17). The wavelengths, λ, of the absorbed light are plotted on the horizontal axis; the vertical axis shows absorbed light intensity.

quantum jump. To do so, the molecule has to leave the empirical world and enter the virtual state space; then it depends on the wave function of the virtual states whether a jump will occur. Thus, the virtual state wave functions are an important factor in the decision of whether a transition is made and how likely it is. *The decision is made before the virtual states are actualized in a transition. Since they affect empirical phenomena, virtual states are real!*

At this point, a simple thought experiment can further illustrate the peculiar nature of the quantum processes. Let's assume that we have been able to build a special spectrometer in which you can see how *single* molecules behave when they are bombarded with a mixture of photons. In this experiment one molecule after another is dismissed, all by itself, into the apparatus, where it is irradiated with light, and then the frequency of the photon that it absorbs is measured. In this setup, it is completely unpredictable what kind of photon or energy quantum the molecule will absorb. Perhaps, using the hypothetical case of figure 18, it will absorb the photon whose energy will propel it from v = 0 and v' = 3. But perhaps it will choose the photon that will allow it to make the transition from v = 0 to v' = 5. The choice is the molecule's choice. Nothing that we know determines this choice. '

As always in the quantum processes, the outcome of a single event is unpredictable, because a single event is ruled by nothing but chance! The results of many events, however—in this case the intensities of the absorptions—are precisely predictable.

ARE VIRTUAL STATES REAL?

Since its conception in the European Renaissance, Western science has been an empirical science. Its procedures are based on the dogma that only statements tested by an experience of the world can be accepted as truths. Thus, taking a physical science—quantum physics—into the context of a transempirical domain of reality does not lead to wild parties, but more typically to angry reactions. Without any doubt, the discovery of an invisible realm of the physical world is a paradox and a challenge, and it has to be accepted with caution. Thus it is necessary to demand that, if virtual states are real, there must be some empirical phenomena that prove that they exist. The question of whether there is a nonempirical part of reality and if virtual states are real has held me in its grip for decades, and in this section I want to summarize for you some of the arguments I have discussed in numerous essays, including analyses worked out with my friends Diogo Valadas Ponte and Sisir Roy and described in *Zygon: Journal of Religion and Science.*

In reacting to the strangeness of the quantum phenomena, Niels Bohr developed the view that, as James Cushing describes, it was "an error of classical realism" to believe that our experience of the visible world can show us what the world is like. Bohr is such a giant figure in the history of quantum theory that his views are still religiously repeated, like prayers, whether they are acceptable or not. Cushing offers a fascinating analysis of Bohr's philosophy and his view that "theory is an abstraction whose components . . . do not represent properties of independent objects (as opposed to the case in classical mechanics)."

In contrast to Bohr's somewhat pessimistic view, I believe that we can make meaningful statements about the nature of reality, but there is a challenge in that we are dealing with a visible reality rooted in a transcendent world that we can't directly experience. In this world, entities exist that are invisible to us, and yet they are real because they have the potential to affect our life. In this regard, science is facing an unavoidable paradox: Even though it must avoid in its descriptions of the world any reference to a transcendent realm of reality, scientific explorations of the world nevertheless force us to accept that such a transcendent realm exists. This finding is of immense significance—as important as the technological, economic, and cultural consequences of the discovery of the quantum phenomena.

In the eighteenth century, Immanuel Kant pointed out that our observation of the world is of the appearance of things, and not of the things as they are in themselves. He called a real thing a *thing in itself,* or a *noumenon.* Noumena, he said, can't be observed. Human knowledge is based on our experiences of the world, but it also contains nonempirical elements, which he called *a priori* because they can't be derived from experience. Thus, there aren't only the facts of our observations of the world; there are also the nonempirical forms into which our observations have to be wrapped so that they can have a meaning in our mind.

Rom Harré has pointed out that Bohr was inspired by Kant's philosophy, and not by the facts of physics, when he proposed that our experience isn't of things, but of our experience of things. This means that our experience can't tell us what things are really like. In discussions by quantum physicists, Bohr's metaphysics has often been translated into the view that the instruments we use for our experiments force a noumenon "to manifest itself in ways that are predetermined by the structure and other properties of the equipment," as Harré describes it. This means that, in contrast to the view that we have developed in this book, electrons can appear to us as waves or particles not because ETs can be real waves or particles, but because we are using instruments that force them to appear in such states. In Rom Harré's description, "the particles . . . can exist nowhere else but in relation to that kind of apparatus." Any information of the world necessarily puts the world into a specific form, like the word *in-form-ation* implies: It is "the putting of something into a form." This difficulty of our knowledge of the world is aggravated even more when we discover that we are also dealing with a nonempirical part of the physical reality.

In this context the nature of potentiality waves is particularly significant. Are they truly existing entities, or are they just concepts of our thinking? The British physicist C. N. Villars was the first to propose that potentiality waves are "physically real waves which exist in their own right, not merely as representations of the behavior of particles." These waves, Villars recognized, aren't mere probability waves that guide the behavior of microphysical objects; "microphysical objects *are* waves of potential observation interactions." With *observation interactions,* Villars denotes the way in which ETs interact with an instrument or their environment. It was Villars's view of potentiality waves that inspired me to think that elementary particles become actually existing waves when they are left alone, and that electrons actually exist as waves when they are a part of an atom or molecule. You will not be too far off the truth if you think that the many atoms and molecules in your body don't just exist as material things, but temporarily also as potentiality waves. And that makes *you* a complex structure and process of potentiality!

The literature abounds with countless studies, very scholarly and impressive, in which the authors try to prove that the wave functions of quantum states aren't real. Asher Peres, for example, a prominent theoretical physicist, published a brilliant mathematical analysis that led him to the conclusion that the wave functions of quantum physics "cannot be an attribute of a physical system." Instead, he proposes, "the Ψ-symbol (the so-called 'state' or 'wave-function') is not an attribute of a *system* but of a *procedure*." He concludes, "A single physical system has no state." Peres's arguments are expert and elegant. If true, they would mean that not only are the wave functions of quantum states not real, but the virtual states themselves aren't either. As it turns out, Peres's views are difficult to reconcile with the empirical facts, particularly with the chemical properties of atoms and molecules.

For example, the chemical reactivity of each individual molecule depends on its quantum states. In particular, the virtual states are an important part of how a molecule can react with others. In seems safe to say that, if easily accessible, empty states wouldn't exist in individual molecules: There would be no chemistry.

Redox (reduction-oxidation) reactions are a special class of chemical reactions in which the reacting atoms and molecules exchange electrons. A metal atom, for example, can lose electrons in a chemical reaction, and then it is said to be *oxidized*. It can also gain electrons, and then it is said to be *reduced*. When electrons are exchanged in this way, empty or virtual states in the receiving atoms become occupied, while occupied states in the donating atoms become virtual. These changes can lead to changes in the magnetic properties of the reacting atoms, which depend on their orbital structures and can be measured. It is hard to see how such reactions could happen or how individual molecules could have magnetic properties if they didn't have any states. We have to conclude that the quantum states and their orbitals, including the virtual states, truly exist in atoms and molecules and are real, because they affect observable chemical reactions and measurable physical properties. The precise prediction of a physical quantity—in this case, the magnetic property of an atom—not only allows us but *forces* us to conclude that the virtual states are elements of reality.

During the past couple of decades or so, the instruments used in molecular spectroscopy have become so powerful that it is now possible to study the photons that *individual* molecules emit. These instruments have opened an important new field of chemical research: *single-molecule spectroscopy*. When a molecule emits a photon, it *must* make a transition from a higher (more energetic) quantum state to a lower one. No other explanation is currently possible. Thus, the emission of photons by single molecules must mean that single molecules exist in quantum states, in contrast to Peres's claim that a "single physical system has no state."

Field-ion microscopes are powerful instruments with which one can image individual atoms. For example, in 2006 Moh'd Rezeq, Jason Pitters, and Robert Wolkow published the image of a tungsten needle in which single atoms are seen. It is difficult to accept the widespread view that these atoms "can exist nowhere else but in relation to that kind of apparatus," as Harré described this school of physics.

Anyone who has been pricked by a needle knows that this isn't the case.

Notes

Preface

xxiii **"A new scientific truth"** Max Planck, *Wissenschaftliche Selbstbiographie*, 22.

xxiv **"the multitude of errors that I had accepted"** Descartes, *Meditationes*, translation by Küng, *Does God Exist?*, 11.

Introduction. Your Cosmic Potential: Being Part of the Universe

3 **"all that, which is hidden behind"** Hüther, *Die Macht der inneren Bilder*, 17.

5 **"all things are number"** Russell, *History of Western Philosophy*, 54.

11 **"the possible to the factual"** Heisenberg, *Physik und Philosophie*, 80.

11 **"tendencies or possibilities"** Ibid., 262.

11 learning process Dürr, *Auch die Wissenschaft spricht nur in Gleichnissen*, 103.

13 **"Reality reveals itself"** Ibid., 12.

14 led us to numerous discoveries of errors Schäfer, "Gradient Revolution in Structural Chemistry."

14 **In the case of proteins . . . we were able to calculate** Jiang et al., "Predictions of Protein Backbone Structural Parameters"; Schäfer et al., "Predictions of Protein Backbone Bond Distances."

15 **"The revelation by modern physics of the void"** . . . **"The atom is as porous as the solar system"** . . . **"If we eliminated"** Eddington, *Nature of the Physical World*, 1.

15 **"The entire cosmos is"** Hirschberger, *Geschichte der Philosophie*, 1:25.

18 **during the past hundred years or so** Bohm, *Wholeness and Implicate Order*, 11; Dürr, *Für eine zivile Gesellschaft*, 18, *Auch die Wissenschaft spricht nur in Gleichnissen*, 102, and *Geist, Kosmos, und Physik*, 44; Eddington, *Nature of the Physical World*, 276, and *Philosophy of Physical Science*, 151; Fischbeck, *Die Wahrheit und das Leben*; Jeans, *Mysterious*

Universe, 158; Kafatos and Nadeau, *Conscious Universe;* Schäfer, *In Search of Divine Reality; Versteckte Wirklichkeit.*

19   **"As a physicist"** . . . **"Basically, there is only spirit!"** Dürr, *Geist, Kosmos, und Physik,* 44, and *Für eine zivile Gesellschaft,* 18.

19   **"The universe is of the nature"** Eddington, *Nature of the Physical World,* 276, and *Philosophy of Physical Science,* 151.

19   **"The universe begins to look"** . . . **"Mind no longer appears"** Jeans, *Mysterious Universe,* 158.

20   **"there is a covenant between our mind"** Schäfer, *In Search of Divine Reality,* 109, and *Versteckte Wirklichkeit,* 159.

22   **"Being means to have been formed"** Hirschberger, *Geschichte der Philosophie,* 1:192.

23   **"The One is all"** Ibid., 1:304.

24   **When Buddhists speak of *Alayavijnana*** Suzuki, *Studies in the Lankavatara Sutra,* 176.

24   **which he called the archetypes** Jung, *Archetypes and the Collective Unconscious,* 4.

24   **"our imagination, perception, and thinking"** Ibid., 44.

24   **typical modes of apprehension** Jung, *Structure and Dynamics of the Psyche,* 137.

24   **psychic organs present in all of us** Jung, *Archetypes and the Collective Unconscious,* 4.

25   *collective unconscious:* **"A psychic system"** Ibid., 43.

25   **"have never been in consciousness"** Ibid., 42.

25   **"a boundless expanse"** . . . **"where I am indivisibly"** . . . **"There I am utterly"** Ibid., 21.

26   **"Religious or metaphysical ways"** Arsac et al., "Pour une science sans a priori."

26   **"the self-motivated Spirit"** Hirschberger, *Geschichte der Philosophie,* 2:412.

27   **"Man knows of god only"** Ibid., 2:419.

27   **"The spirit of human beings"** Ibid.

27   **"The truth is the whole"** Ibid., 2:412.

27   **"undivided wholeness"** Bohm, *Wholeness and Implicate Order,* 11.

27   **the argument that physicist Menas Kafatos and science historian Robert Nadeau used** Kafatos and Nadeau, *Conscious Universe.*

27   **"Consciousness amounts to"** Lancaster, *Approaches to Consciousness,* 91.

28   **"My experiences from psychosomatic"** Personal communication with Hanne Seemann, March 11, 2009.

28   **"Often when I wake up"** Buber, "Ekstatische Konfessionen," 88.

28   **Gospel of Thomas we read in logion 117** Shoemaker, "English Translation of the Gospel of Thomas."

29   **philosopher Jean Gebser has described** Gebser, *Ursprung und Gegenwart.*

29   **"Like the meridians"** Teilhard, *Phenomenon of Man,* 30.

30   **"God is an unspoken word"** Eckhart, *Werke 1.*

Chapter 1. Materialism Is Wrong: The Basis of the Material World Is Nonmaterial

33   **"Modern atomic theory is thus essentially different"** Heisenberg, "Ideas of the Natural Philosophy," 55.

34   **"God in the beginning formed Matter"** Newton, *Opticks,* 400.

38   **waves come *legato*** Polkinghorne, 1984.

47   **quantum waves are simply some tool** Bohr, *Philosophical Writings of Niels Bohr.*

47   **In some of his writings** Heisenberg, "Ideas of the Natural Philosophy," and *Physics and Philosophy.*

47   **particles and waves are real** Bohm, *Wholeness and Implicate Order.*

47   ***potentiality* waves** Villars, "Observables, States, and Measurements in Quantum Physics," and "Microphysical Objects as 'Potentiality Waves.'"

48   **potentiality waves "are conceived as"** Villars, "Microphysical Objects as 'Potentiality Waves," 148.

48   **physical properties . . . don't have an *actual* value . . . it doesn't exist . . . "in ordinary three-dimensional space"** Ibid.

48   **"During the act of observation"** Heisenberg, *Physik und Philosophie,* 80.

Chapter 2. Your Potential Is Real Even Though You Can't See It: How Virtual States Act On the Visible World

54   **"Not all potentiality is converted to actuality"** Wheeler and Ford, *Geons, Black Holes and Quantum Foam,* 338.

61   **"Atoms do not exist as"** Heisenberg, "Ideas of the Natural Philosophy," 56.

74   **"Modern atomic theory" . . . "Atoms are no longer"** Ibid., 55.

CHAPTER 3. WE ARE ALL CONNECTED: REALITY AS INDIVISIBLE WHOLENESS

75   "Relativity and quantum theory agree, in that they both imply" Bohm, *Wholeness and Implicate Order,* 11.

77   "total order" . . . "is contained" . . . "total structure 'enfolded'" Bohm, *Wholeness and Implicate Order,* 149.

77   "unbroken" . . . "undivided wholeness" . . . Out of the constantly changing . . . "relevate" . . . "mind and matter are not" Ibid., 11, 151.

78   As Deepak Chopra explains Chopra, "Meditation Techniques."

79   "a guideline for research" Weinberg, *Dreams of a Final Theory,* 52, 54.

80   "At the other end of the spectrum" . . . "To whatever extent" Ibid., 52.

81   "At its nuttiest extreme" . . . "The reductionist worldview is" Ibid, 53.

83   "implicate order" Bohm, *Wholeness and Implicate Order.*

83   "is in instant contact" Herbert, *Quantum Reality,* 188.

83   "Quantum physics has revealed" Dürr and Oesterreicher, *Wir erleben mehr als wir begreifen,* 21.

83   "When the ocean is" Dürr, *Auch die Wissenschaft spricht nur in Gleichnissen,* 102.

84   "reality reveals itself" Ibid., 12.

84   "Potentiality appears as" Ibid.

86   "coincidence of matter and form" Bruno, *Cause, Principle, and Unity,* 10.

86   "For you must know that" Edwards, *Encyclopedia of Philosophy,* 1:407.

87   "The universe is all in one" Hirschberger, *Geschichte der Philosophie,* 2:39.

87   Dürr made the connection Dürr, *Auch die Wissenschaft spricht nur in Gleichnissen,* 102.

89   "respond together to" Kafatos and Nadeau, *Conscious Universe,* 73.

89   "even if they occupy" D'Espagnat, *In Search of Reality,* 45.

90   "that nonseparability is" Ibid., 46.

90   "infer" . . . don't "prove" . . . "the reality that exists" Kafatos and Nadeau, *Conscious Universe,* 9.

90   "All that we can say" Ibid.

91   "a fundamental property" . . . "factual condition" Ibid., 9, 73.

92   "humanistic psychology" Maslow, "Theory of Human Motivation," *Religions, Values, and Peak Experiences, Toward a Psychology of Being,* and *Farther Reaches of Human Nature.*

92   the steps of a "pyramid" Maslow, *Toward a Psychology of Being,* 168, and "Theory of Human Motivation."

93   "A musician must make" . . . need of "self-actualization" . . . "refers to the desire" Maslow, "Theory of Human Motivation," 382.

Chapter 4. Consciousness: A Cosmic Property

95   "The teaching of Sri Aurobindo starts from that of the ancient sages" Aurobindo, *Writings of Sri Aurobindo,* 39.

98   "But the word is neither" . . . "you have heard" . . . "Like my word" Augustine, Sermon CCXXV.3, translated from Latin by Erich F. Paulus in a personal communication, 2003.

99   "your number came up" Monod, *Chance and Necessity.*

100   "Before I formed you" Jeremiah 1:5; New Revised Standard Version; I owe this citation to one of my students, Lacy Fincannon.

100   "that it leaves us" Eddington, *Nature of the Physical World,* 309.

101   In the books of physicists Dürr, *Für eine zivile Gesellschaft* and *Auch die Wissenschaft spricht nur in Gleichnissen;* Fischbeck, *Die Wahrheit und das Leben.*

102   "The universe is of the nature" Eddington, *Philosophy of Physical Science,* 151.

102   "Now we realize" Eddington, *Nature of the Physical World,* 259.

102   "why not then attach" Ibid.

103   "there is a background" Ibid., 312.

103   "If the unity of" Ibid., 315.

104   "has a nature capable of" . . . "the more specialized" Ibid., 259, 260.

104   "the stuff of the world" . . . "The mind-stuff of the world" Ibid., 276.

105   "It is difficult for" Ibid., 281.

105   "We have only one approach" . . . "Consciousness is not sharply" . . . "to our conscious feelings" Ibid., 280.

106   "'implies' without being" Kafatos and Nadeau, *Conscious Universe,* 109.

106   a condition of our mind Ibid., 129.

106   the universe is conscious Ibid., 170.

109   called *object permanence* by the Swiss psychologist Piaget, *Child's Construction of Reality.*

110   "No object ever discovers" David Hume, *Enquiry Concerning Human Understanding,* IV.1.

111   "meaningful coincidence" . . . "something other than the probability

of chance" . . . "takes place outside" . . . "No one has yet succeeded" . . . "based on some kind of principle" . . . "the common bridge" Jung, *Structure and Dynamics of the Psyche,* 520–31.

113   "the psyche cannot" Ibid., 531.

114   "explanatory physics, where" Omnès, *Quantum Philosophy,* 45.

114   "evocative" . . . "reproductive" . . . "ground of existence" Haftmann, *Painting in the Twentieth Century.*

115   "excavation of the unconscious" Gebser, *Ursprung und Gegenwart,* 171.

118   "cold and warm, dry and wet" Störig, *Kleine Weltgeschichte der Philosophy,* 111.

118   Hirschberger has discussed how . . . must be infinite Hirschberger, *Geschichte der Philosophie,* 1:20.

118   "Constantly new worlds are emerging" Störig, *Kleine Weltgeschichte der Philosophy,* 111.

118   *spanda* . . . "vibrations in the divine" Muller-Ortega, *Triadic Heart of Śiva,* chap. 6.

119   "That self-consciousness in the Heart" . . . *spanda* "is the wave of the ocean" Ibid., 118.

119   "The Ultimate" . . . "internal dynamism" . . . "results in the process of manifestation" . . . "full of bliss" Ibid., 121.

120   "The silence of the Supreme" . . . the "vibration that characterizes consciousness" Ibid., 120.

122   "but on a 'mythological' conception" Jung, *Archetypes and the Collective Unconscious,* 57.

CHAPTER 5. DARWIN WAS WRONG: EVOLUTION NEEDS QUANTUM SELECTION AND COOPERATION

127   "It must be recognized that monist-materialism leads" Eccles, *The Human Mystery,* 9.

128   "total solitude," in "fundamental isolation," and "on the boundary" . . . "like a gypsy" Monod, *Chance and Necessity,* 160.

129   "with the material" Miller, *Finding Darwin's God,* 184.

129   "a form of practical" Ibid., 194.

129   "all that Darwin did" Ibid., 168.

130   quantum chemical calculations of the structures of peptide Schäfer, Van Alsenoy, and Scarsdale, "Ab Initio Studies of Structural Features."

130   **calculations were found to be in perfect agreement** Jiang et al., "Predictions of Protein Backbone Structural Parameters."

130   **"contingent assemblages of matter"** Denton, Marshall, and Legge, "Protein Folds as Platonic Forms," 334.

131   **As many of our calculations showed** See the references in Jiang et al., "Predictions of Protein Backbone Structural Parameters."

131   **it is practically impossible** Denton, Marshall, and Legge, "Protein Folds as Platonic Forms," 337–41.

131   **"Science works because"** . . . **"Once you understand"** Miller, *Finding Darwin's God,* 194.

131   **"mechanistic sum"** Ibid., 194.

132   *nature doesn't make jumps* Darwin, *Origin of Species,* 203.

133   **"The question of whether"** . . . **"no general bias"** . . . **"Mutations are caused"** Dawkins, *Blind Watchmaker,* 306.

135   **"the minutest phenomena of nature"** . . . **"we have, for instance"** Jeans, *Mysterious Universe,* 132.

136   **a "choice" is made "on the part of nature"** Dirac, cited by Stapp, *Mind, Matter, and Quantum Mechanics,* 190.

136   **"a 'choice' is defined"** Stapp, *Mind, Matter, and Quantum Mechanics,* 195.

137   **If gene molecules were** Denton, Marshall, and Legge, "Protein Folds as Platonic Forms," 334.

139   **"on the other, all"** Bauer, *Das Kooperative Gen,* 72.

139   **"Creating something from nothing"** Deacon, "Three Levels of Emergent Phenomena," 8.

142   **"Facts from science tell us"** Pollack, "Unknown, the Unknowable, and Free Will," 7.

142   **"A *totally* blind process"** Monod, *Chance and Necessity,* 96.

144   **"the psychic" and of "thought"** Teilhard, *Phenomenon of Man,* 30.

144   **"noise" that natural selection** . . . **"music"** Monod, *Chance and Necessity,* 113.

145   **possible by biological cooperation** Bauer, *Das Kooperative Gen;* Joseph, "Extinction, Metamorphosis, Evolutionary Apoptosis"; Lipton, *Biology of Belief;* Woese, "On the Evolution of Cells."

145   **Genes aren't "selfish"** Dawkins, *Selfish Gene.*

145   **"cooperativeness, communication, and creativity"** Bauer, *Das Kooperative Gen,* 17.

146   **Richard Dawkins's account** Dawkins, *Selfish Gene.*

146   **Rhawn Joseph, neurobiologist Joachim Bauer, and microbiologist**

Carl Woese Joseph, *"Extinction, Metamorphosis, Evolutionary Apopto-sis"; Bauer, Das Kooperative Gen;* Woese, *"On the Evolution of Cells,"* 8742.

146 **"primeval soup"** ... **"replicator molecules"** ... **"were the ancestors of life"** Dawkins, *Selfish Gene,* 18.

146 **"There was a struggle"** ... **"Any mis-copying"** Dawkins, *Selfish Gene,* 19.

146 **which *shared* information** Joseph, "Extinction, Metamorphosis, Evolutionary Apoptosis."

146 **"It is the community"** Woese, "On the Evolution of Cells," 8742.

146 **willingness to cooperate** Joseph, "Extinction, Metamorphosis, Evolutionary Apoptosis"; Bauer, *Das Kooperative Gen.*

147 **"lone warriors"** ... **"of biological systems"** Bauer, *Das Kooperative Gen,* 54.

147 **"optimized strategies"** Ibid., 15.

147 **"is said to be altruistic"** Dawkins, *Extended Phenotype,* 291.

147 **contradicted by recent research** Joseph, "Extinction, Metamorphosis, Evolutionary Apoptosis"; Bauer, *Das Kooperative Gen.*

147 **"cells may have mechanisms"** Cairns, Overbaugh, and Miller, "Origin of Mutants," 142.

147 **"surges"** Bauer, *Das Kooperative Gen,* 16; Gould, *Structure of Evolutionary Theory.*

147 **"the self-modification of"** Bauer, *Das Kooperative Gen,* 66.

147 **"Species formations are the work"** Ibid., 188.

148 **In an exciting paper** Joseph, "Extinction, Metamorphosis, Evolutionary Apoptosis."

148 **"microbes to arrive"** ... **"altering the womb"** Ibid.

149 **" 'evolution' *is under genetic regulatory control*"** ... **"contained the genes"** ... **"genetic seeds"** Ibid.

149 **"Like programmed cell death"** Ibid.

150 **"just a sample"** Ibid.

150 **"the boundary of an alien world"** Monod, *Chance and Necessity.*

151 **"Machines cannot receive"** Personal communication with Karl Goser, September 2005; Goser, "Von der Information zur Transzendenz."

151 **the pre-Darwinian biologists** Owen, *On the Nature of Limbs,* and *Anatomy of Vertebrates.*

151 **"unity which underlies"** Owen, *Anatomy of Vertebrates,* v.

151 **Laws of Form** See Denton, Marshall, and Legge, "Protein Folds as Platonic Forms," 326.

152 "archetypes" . . . "from the general" Owen, *Anatomy of Vertebrates,* xxv and vi.

152 considered form primary Denton, Marshall, and Legge, "Protein Folds as Platonic Forms," 326.

152 "necessity was replaced" Ibid., 328.

153 "vehicle" for their selfish purposes Dawkins, *Extended Phenotype,* 4, 302.

153 "They are in you" . . . "They have come a long way" Dawkins, *Selfish Gene,* 20.

154 "is merely an adaptation" . . . "the basis of ethics" Ruse and Wilson, "Approach of Sociobiology," 310.

155 "The way our biology enforces" Ibid., 311.

155 "Our morality is put in place" Ruse, *The Bad Smell of Anti-Reductionism.*

155 "a collective illusion of humankind" Ruse, *The Confessions of a Skeptic.*

156 "the good of the race" . . . "Evolutionary Ethics" Weikart, *From Darwin to Hitler.*

157 "We can 'infer'" Kafatos and Nadeau, *Conscious Universe,* 177.

157 the numerous suggestions Bohm, *Wholeness and Implicate Order;* Dürr, *Für eine zivile Gesellschaft;* Eddington, *Nature of the Physical World,* and *Philosophy of Physical Science;* Fischbeck, *Die Wahrheit und das Leben;* Jeans, *Mysterious Universe;* Lipton, *Biology of Belief;* Kafatos and Nadeau, *Conscious Universe;* Schäfer, *In Search of Divine Reality;* Versteckte Wirklichkeit.*

157 "consciousness amounts to" Lancaster, *Approaches to Consciousness,* 91.

158 this leads me to think Schäfer, "Quantum Reality and Evolution Theory" and *"Paraklase der Weltsicht."*

159 "Each cell is" Lipton, *Biology of Belief,* 7.

160 cooperation, not competition Dürr, *Für eine zivile Gesellschaft,* 29.

160 "We are not 'stuck' with" Lipton, *Biology of Belief,* 169.

160 "survival of the most loving" Ibid., 171.

161 human beings aren't wired Bauer, *Prinzip Menschlichkeit.*

161 "The core of all motivation" Ibid., 36.

161 "psychology into biology" Ibid., 8.

162 "to burst from the smallest crack" . . . "once it appeared, it" Teilhard, *La Place de l'Homme Dans La Nature,* 50.

CHAPTER 6. WORLD ETHOS: LIVING IN HARMONY WITH THE ORDER OF THE UNIVERSE

164 **"Our natures are parts of the World-Whole"** Zeno of Citium, see Hauskeller, *Geschichte der Ethik*, 204

165 **My discussion of these issues** Schäfer, Valadas Ponte, and Roy, "Quantum Reality and Ethos" and "Quantenwirklichkeit und Weltethos."

165 **This exchange of ideas significantly expanded** Schäfer, *In Search of Divine Reality;* "Quantum Reality, the Emergence of Complex Order from Virtual States, and the Importance of Consciousness in the Universe"; "Response to Erwin Laszlo: Quantum and Consciousness"; "A Response to Carl Helrich: The Limitations and Promise of Quantum Theory"; and "A Response to Stanley Klein: A Dialogue on the Relevance of Quantum Theory to Religion."

167 **"naturalistic fallacy"** Moore, *Principia Ethica.*

168 **"to live in accordance with Nature"** Hauskeller, *Geschichte der Ethik*, 203.

168 **"value-best state"** Ibid., 21.

169 **"world-principle which in human beings"** . . . **"harmony with nature"** . . . **"Our natures are parts"** . . . **"Therefore, the final goal"** . . . **"one undertakes nothing"** . . . **"nature of the all-pervading"** Ibid., 202–4.

170 **that appeared in my own mind** Schäfer, *In Search of Divine Reality.*

170 **"of cosmic morality and hope"** Klein, "A Dialogue on the Relevance of Quantum Theory," 569.

172 **"leads a robber's life"** Plato, *Gorgias, 508a.* See Platon, *Sämtliche Werke;* or Plato, *The Essential Dialogues of Plato.*

173 **See the inspiring books** Dalai Lama and Cutler, *Art of Happiness;* Ricard, *Happiness.*

173 *eudaimonia* **is often translated as . . . fulfillment of a successful life** Hauskeller, *Geschichte der Ethik,* 83.

173 **"Pleasure signals to a sentient being"** Ibid., 142.

173 **"Among everything that wisdom"** Hirschberger, *Geschichte der Philosophie,* 1:288.

173 **In the metaphysics of Immanuel Kant . . . The highest good . . . "sufficient connection of happiness"** Kant, *Kritik der praktischen Vernunft,* 177–82.

174 **In our approach** Schäfer, Valadas Ponte, and Roy, "Quantum Reality and Ethos," and "Quantenwirklichkeit und Weltethos."

174 **Bruce Lipton describes** Lipton, *Biology of Belief,* 116.

174   **human beings aren't laid out** You will find an excellent description in Joachim Bauer, *Prinzip Menschlichkeit*.

174   **"To find and to give"** Ibid., 36.

174   **Since your brain "turns psychology"** Ibid., 8.

176   **"inner sense of existence,"** Biran, *L'Effort, Textes choisis par A. Drevet*.

176   **"as identified with *me*"** Ibid., 115.

176   **"the fact of existence"** Ibid., 144.

177   **Even though it is used** Jung, *Structure and Dynamics of the Psyche*, 233.

179   **We can call this the principle of "self-permanence"** Schäfer, Valadas Ponte, and Roy, "Quantum Reality and Ethos," and "Quantenwirklichkeit und Weltethos."

179   **Not only can't you observe** Schäfer, *Versteckte Wirklichkeit*, 235.

180   **"Strictly speaking"** Dürr, *Für eine zivile Gesellschaft*, 63.

180   **"there is no being"** Ibid., 18.

183   **in the sense of *aretē*** Hauskeller, *Geschichte der Ethik*, 21.

183   **Inside of our consciousness... "the subjective feeling of freedom"... "Just as our free will clashes"... "like an *objective occurrence*"** Jung, *Archetypes and the Collective Unconscious*, 5, 6.

184   **how European thinking evolved** Küng, *Does God Exist?*, 91.

185   **how La Mettrie concluded** Ibid.

185   **In his book *System of Nature*... he declares, as Küng describes it, "matter and mind"... "priests must be replaced"** Ibid., 92.

185   **"monist's perspective"... "brain and the mind are inseparable events"... "a living entity"** Llinás, *I of the Vortex*, 1, 2.

185   **"Mindness coincides with functional"** Ibid., 3

186   **"The only reality that exists"** Ibid., 259.

186   **that feeling is an illusion** Ibid., 127.

186   **"Self" or "I"... a "tangible thing"... a "useful construct"... "Uncle Sam"... a "particular mental state"... "a calculated entity"** Ibid., 127–28.

186   **"Self," Llinás writes, "is the invention of an intrinsic"** Ibid., 128.

186   **"So, now we have a wondrous"** Ibid., 133.

187   **what Diogo Valadas Ponte, Sisir Roy, and I have called *tacit advice*** Schäfer, Valadas Ponte, and Roy, "Quantum Reality and Ethos," and "Quantenwirklichkeit und Weltethos."

188   **"We do not know"** Jung, *Aion*, 26.

188   **"The awareness of the moral law"** Kant, *Kritik der praktischen Vernunft*, 55.

188   **"All moral concepts have their seat"** Kant, *Groundwork of the Meta-physics of Morals*, 79.

188   **moral principles "cannot be abstracted"** Ibid.

188   **its reliance on a *single* principle** Schäfer, Valadas Ponte, and Roy, "Quantum Reality and Ethos," and "Quantenwirklichkeit und Weltethos."

189   **All explicit formulations of moral laws** Ibid.

190   **In his *Protagoras*, Plato considered** Plato, *Protagoras 329d,* See Platon, *Sämtliche Werke*; or Plato, *The Essential Dialogues of Plato.*

190   **our view is different because it assumes** Schäfer, Valadas Ponte, and Roy, "Quantum Reality and Ethos," and "Quantenwirklichkeit und Weltethos."

190   **C. N. Villars's description of potentiality waves** Villars, "Microphysical Objects as 'Potentiality Waves.'"

190   **Kant thought that "moral reason is"** Hirschberger, *Geschichte der Philosophie*, 2:338–40.

190   **who noted this aspect of Christian love** Weizsäcker, *Die Sterne sind glühende Gaskugeln*, 30.

190   **Bernardin Schellenberger described** Schellenberger, *Entdecke dass du glücklich bist.*

190   **"does not speak of duties" . . . The beatitudes "are no obligations"** Ibid., 15, 17.

191   **Their messages depend** Jung [1959] 1990, 5.

192   **"does not come from" . . . "One has only to think"** Jung, *Psychology and Religion*, 197.

192   **presented arguments for the thesis** Schäfer, Valadas Ponte, and Roy, "Quantum Reality and Ethos," and "Quantenwirklichkeit und Weltethos."

192   **arguments for the existence of a cosmic memory field** Dürr, *Auch die Wissenschaft spricht nur in Gleichnissen,* 67; Laszlo, *Science and the Akashic Field,* 75.

192   **"menacing power that lies fettered"** Jung, *Structure and Dynamics of the Psyche,* 206.

192   **"everything is liable" . . . "the unconscious may reveal"** Jung, *Mysterium Coniunctionis*, 367–68.

193   **"on the brink of actions" . . . "mankind is powerless"** Jung, *Archetypes and the Collective Unconscious,* 23.

193   **Socrates already taught** Plato, *Meno,* 87c–89a, See Platon, *Sämtliche Werke*; or Plato, *The Essential Dialogues of Plato.*

193   **"Nobody commits an evil act"** Plato, *Protagoras,* 345e, 355c, See Plato, *Sämtliche Werke*; or Plato, *The Essential Dialogues of Plato.*

193   **"The only thing that is good"** Kant, *Groundwork of the Metaphysics of Morals,* 17.

193   **"The far goal of education"** Maslow, *Religions, Values, and Peak Experiences,* 49.

194   **Contemporary psychology shows us** Neumann, *Origins and History of Consciousness;* Kohlberg, Levine, and Hewer, *Moral Stages.*

194   **"Although contemporary man believes"** Jung, *Mysterium Coniunctionis,* xviii.

Chapter 7. Your Evolving Mind: Integrative Consciousness and a Leap into a New Human Species

195   **"For thirty years I went about searching for God"** Bâjewîd Bestâmi; see Martin Buber, "Ekstatische Konfessionen," 67.

196   **Jean Gebser's thesis . . . another mutation is impending** Gebser, *Ursprung und Gegenwart.*

198   **the archetypes are nonempirical** Jung, *Archetypes and the Collective Unconscious,* 42.

199   **Understanding the past . . . Jean Gebser thought** Gebser, *Ursprung und Gegenwart.*

200   **an evolving structure is reached in an unpredictable "jump"** Ibid., 166.

200   **"reconstruct the human 'factor'" . . . "restoration of the unharmed" . . . "were not distinguished"** Ibid., 83.

200   **the newly emerging integrative consciousness** Ibid., 167.

201   **Joachim Faulstich has illustrated** Faulstich, *Das Geheimnis der Heilung.*

202   **This is, as Faulstich explains, where Gebser's** Ibid., 12.

202   **"respect the rational component" . . . "cruel rationality"** Ibid., 12.

202   **"For rational reasons"** Ibid., 14.

202   **"Pioneers of an all inclusive"** Ibid., 13.

203   **Jacques Monod was amazed** Monod, *Chance and Necessity.*

205   **Monod calls this the "animist tradition"** Ibid., 38, 163.

205   **"the only source of real truth"** Ibid., 158.

206   **"Every living being is"** Ibid., 150.

206   **"sickness of spirit"** Ibid., 154.

206   **"We would like to think"** Ibid., 50.

206   "epistemological contradiction," Ibid., 30.

206   "to preserve and reproduce" Ibid.

206   "autonomous morphogenesis" Ibid., 26.

207   "objectivity nevertheless obliges" Ibid., 31.

207   "the central problem of biology" Ibid., 31.

209   "virtual being" ... "In its entirety it exists" ... "In its founda-
       tion" ... "The things ... are out of the oneness" ... "In God, how-
       ever" ... the "standing in the oneness" ... "truth of God" ... "can
       never comprehend their" ... "The 'esse virtuale,'" or the *virtual
       being* of things ... "the 'virtue' in the things" ... actualize "in the
       earthly order" ... Kopper, *Die Metaphysik Meister Eckharts,* 63–64.

212   Gerald Hüther believes that the power Hüther, *Die Macht der inneren
       Bilder,* 10.

213   the "cunning of ideas" Hegel; see Hirschberger, *Geschichte der Phi-
       losophie,* 1:99.

214   "The eternal silence of ... I do not know who ... I see the terrify-
       ing spaces ... man's wretchedness ... infinity on every side ... All
       I know is that ... The last act is bloody" Blaise Pascal, citation by
       Küng, *Does God exist?,* 52–55.

215   James S. Cutsinger reports that, when a disciple asked Jakob
       Böhme ... "Son, when you can throw yourself" ... "It is in you, my
       son" ... "God hears and sees through you" Cutsinger, *Not of This
       World,* 5.

216   Martin Buber reports that when Bâjewîd Bestâmi see Buber,
       "Ekstatische Konfessionen," edited by Sloterdijk, "*Mystische Welt-
       literatur,* 67.

APPENDIX FOR CHAPTER 1: ON SINGLE-PARTICLE INTERFERENCE
AND THE CONCEPT OF POTENTIALITY WAVES

230   chemists use to study the structures Schäfer, "Electron Diffraction as
       a Tool of Structural Chemistry," 1976.

234   interference phenomena ... wave properties of neutrons Gähler and
       Zeilinger, "Wave-Optical Experiments with Very Cold Neutrons."

235   helium atoms Pfau et al., "Loss of Spatial Coherence."

235   those of rubidium Dürr, Nonn, and Rempe, "Origin of Quantum-
       Mechanical Complementarity."

235   molecule containing sixty carbon atoms Arndt et al., "Wave-Particle
       Duality of C60 Molecules."

APPENDIX FOR CHAPTER 2: HOW THE NONEMPIRICAL PART OF REALITY IS DISCOVERED IN THE VIRTUAL STATES OF ATOMS AND MOLECULES

247   **In a beautiful study, Wolfgang Rueckner and Paul Titcomb have shown** Rueckner and Titcomb, "A Lecture Demonstration of Single Photon Interference."

259   **"between the idea of an event and"** Heisenberg, *Physics and Philosophy,* 41.

266   **I want to summarize for you** Schäfer, *In Search of Divine Reality, Versteckte Wirklichkeit,* "Die Quantenwirklichkeit als Manifestation," "Quantum Reality, the Emergence of Complex Order," "Paraklase der Weltsicht," and "Nonempirical Reality"; Schäfer, Valadas Ponte, and Roy, "Quantum Reality and Ethos," and "Quantenwirklichkeit und Weltethos."

267   **"an error of classical realism"** Cushing, "Fundamental Problems in and Methodological Lessons," 29.

267   **"theory is an abstraction whose components"** Ibid., 30.

267   **Bohr was inspired by Kant's** Harré, "Parsing the Amplitudes," 66.

268   **"to manifest itself in ways"** . . . **"the particles . . . can exist nowhere"** Ibid.

268   **"physically real waves"** . . . **"microphysical objects *are*"** Villars, "Microphysical Objects as 'Potentiality Waves,'"148.

269   **"cannot be an attribute"** . . . **"the Ψ-symbol"** . . . **"A single physical system"** Peres, "What Is a State Vector?," 646.

270   **the image of a tungsten needle** Rezeq, Pitters, and Wolkow, "Tungsten Nanotip Fabrication."

270   **"can exist nowhere else but"** Harré, "Parsing the Amplitudes," 66.

Sources

Books

Aurobindo, Sri, *Writings of Sri Aurobindo*. Edited by Robert McDermott. Great Barrington, Mass.: Lindisfarne Books, [1987] 2001.

Bauer, Joachim. *Das Kooperative Gen*. Hamburg, Germany: Hoffmann und Campe, 2008.

———. *Prinzip Menschlichkeit*. Hamburg, Germany: Hoffmann und Campe, 2006.

Biran, Marie-François-Pierre Gonthier Maine de. *L'Effort, Textes choisis par A. Drevet*. Presses Universitaires de France, Paris, 1966.

Bohm, David. *Wholeness and Implicate Order*. London: Routledge and Kegan Paul, 1981. First published in 1980.

Bohr, Niels. *The Philosophical Writings of Niels Bohr*. Woodbridge, Conn.: Ox Bow Press, 1987.

Bruno, Giordano. *Cause, Principle, and Unity*. Edited and translated by Robert de Lucca. Cambridge: Cambridge University Press, 1998.

Buber, Martin. "Ekstatische Konfessionen." In *Mystische Weltliteratur*, edited by Peter Sloterdijk. Jena, Germany: Eugen Diederichs, 2007. First published in 1909.

Cutsinger, James S. *Not of This World: A Treasury of Christian Mysticism*. Bloomington, Ind: World Wisdom, 2003.

Cushing, James T. "Fundamental Problems in and Methodological Lessons from Quantum Field Theory." In *Philosophical Foundations of Quantum Field Theory*, edited by Harvey R. Brown and Rom Harré, 25–39. Oxford: Oxford University Press, 2003. First published in 1988.

Dalai Lama and Howard C. Cutler. *The Art of Happiness*. London: Penguin Books, 2009.

Darwin, Charles. *The Origin of Species*. London: Oxford University Press, 1956. First published in 1872.

Dawkins, Richard. *The Blind Watchmaker.* New York: Norton, 1996. First published in 1986.

―――. *The Extended Phenotype.* Oxford: Oxford University Press, 1999. First published in 1982.

―――. *The Selfish Gene.* Oxford: Oxford University Press, 1999. First published in 1976.

D'Espagnat, Bernard. *In Search of Reality.* New York: Springer, 1983. First published in 1979.

Dürr, Hans-Peter. *Auch die Wissenschaft spricht nur in Gleichnissen.* Freiburg, Germany: Herder, 2004.

―――. *Für eine zivile Gesellschaft.* Munich: Deutscher Taschenbuch Verlag, 2000.

―――. *Geist, Kosmos, und Physik.* Amerang, Germany: Crotona Verlag, 2010.

Dürr, Hans-Peter, and M. Oesterreicher. *Wir erleben mehr als wir begreifen.* Freiburg, Germany: Herder, 2001.

Eccles, John C. *The Human Mystery.* New York: Springer International, 1979 (and London, Routledge and Kegan Paul, 1984).

Eckhart, Meister. *Werke 1: Texte und Übersetzungen.* Edited by Josef Quint. Annotated by Niklaus Largier. Frankfurt: Deutscher Klassiker Verlag, 2008.

Eddington, Arthur S. *The Nature of the Physical World.* New York: Macmillan, 1929.

―――. *The Philosophy of Physical Science.* New York: Macmillan, 1939.

Edwards, Paul, ed. *The Encyclopedia of Philosophy.* 8 vols. New York: Macmillan, 1967.

Faulstich, Joachim. *Das Geheimnis der Heilung.* Munich: Knaur, 2010.

Feuerstein, George, Subhash Kak, and David Frawley. *In Search of the Cradle of Civilization.* Wheaton, Ill.: Theosophical Publishing House, 2001. First published in 1995.

Fischbeck, Hans-Jürgen. *Die Wahrheit und das Leben—Wissenschaft und Glaube im 21. Jahrhundert.* Munich: Utz Verlag, 2005.

Gebser, Jean. *Ursprung und Gegenwart.* Quern-Neukirchen, Germany: Novalis Verlag, 1986.

Goser, K. "Von der Information zur Transzendenz—vom Wissen zum Glauben." In *Glaube und Denken. Jahrbuch der Karl-Heim-Gesellschaft*

20. Jahrgang 2007, edited by R. Martin and U. Beuttler, 177–96. Frankfurt am Main: Peter Lang GmbH, 2007.

Gould, S. J. *The Structure of Evolutionary Theory.* Boston: Belknap Press of Harvard University Press, 2002.

Haftmann, Werner. *Painting in the Twentieth Century.* 2 vols. New York: Praeger, 1965.

Harré, Rom. "Parsing the Amplitudes." In *Philosophical Foundations of Quantum Field Theory,* edited by Harvey R. Brown and Rom Harré, 59–71. Oxford: Oxford University Press, 2003. First published in 1988.

Hauskeller, Michael. *Geschichte der Ethik: Antike* (History of Ethics: The Era of the Greeks). Munich: Deutscher Taschenbuch Verlag, 1997.

Heisenberg, Werner. "Ideas of the Natural Philosophy of Ancient Times in Modern Physics." In *Philosophical Problems of Quantum Physics,* 53–59. Woodbridge, Conn.: Ox Bow, 1979. First published in 1952.

———. *Physics and Philosophy.* New York: Harper Torchbooks, 1962. First published in 1958.

———. *Physik und Philosophie.* Stuttgart, Germany: Hirzel, 2000. First published in 1958.

Herbert, Nick. *Quantum Reality.* Garden City, N.Y.: Anchor Press, 1985.

Hirschberger, Johannes. *Geschichte der Philosophie.* 2 vols. Freiburg, Germany: Herder, 1981. First published in 1965.

Hüther, Gerald. *Die Macht der inneren Bilder.* Göttingen, Germany: Vandenhoeck und Ruprecht, 2010.

James, William. *Varieties of Religious Experience.* Charleston, S.C.: Biblio Bazaar, 2007. First published in 1902.

Jeans, James. *The Mysterious Universe.* New York: Macmillan, 1931.

Jung, Carl Gustav. *Aion.* Princeton, N.J.: Princeton University Press, 1978. First published in 1959.

———. *The Archetypes and the Collective Unconscious.* Princeton, N.J.: Princeton University Press, 1990. First published in 1959.

———. *Psychology and Religion: West and East.* Princeton, N.J.: Princeton University Press, 1989. First published in 1958.

———. *Mysterium Coniunctionis.* Princeton, N.J.: Princeton University Press, 1989. First published in 1963.

———. *The Red Book.* New York: W. W. Norton and Company, 2009.

————. *The Structure and Dynamics of the Psyche.* Princeton, N.J.: Princeton University Press, 1981. First published in 1960.

Kafatos, Menas, and Robert Nadeau. *The Conscious Universe.* New York: Springer, 1990.

Kant, Immanuel. *Groundwork of the Metaphysics of Morals.* New York: Harper Torchbooks, 1964. First published in 1785.

————. *Kritik der praktischen Vernunft.* Stuttgart, Germany: Reclam, 1998. First published in 1788.

Kohlberg, Lawrence, Charles Levine, and Alexandra Hewer. *Moral Stages: A Current Formulation and a Response to Critics.* New York: S. Karger, 1983.

Kopper, Joachim. *Die Metaphysik Meister Eckharts.* Saarbrücken, Germany: West-Ost-Verlag, 1955.

Küng, Hans. *Does God Exist?* Garden City, N.Y.: Doubleday, 1980.

Lancaster, Brian L. *Approaches to Consciousness.* New York: Palgrave Macmillan, 2004.

Laszlo, Ervin. *Science and the Akashic Field.* Rochester, Vt.: Inner Traditions, 2007. First published in 2004.

Lipton, Bruce H. *The Biology of Belief.* New York: Hay House, 2005.

Llinás, Rodolfo R. *I of the Vortex: From Neurons to Self.* Cambridge, Mass.: MIT Press, 2002.

Maslow, Abraham H. *The Farther Reaches of Human Nature.* New York: Penguin, 1993. First published in 1971.

————. *Religions, Values, and Peak Experiences.* New York: Penguin, 1994. First published in 1964.

————. *Toward a Psychology of Being.* New York: John Wiley, 1999. First published in 1968.

Miller, Kenneth R. *Finding Darwin's God.* New York: Cliff Street Books, 1999.

Monod, Jacques. *Chance and Necessity.* London: Collins, 1972.

Moore, George Edward. *Principia Ethica.* Cambridge: Cambridge University Press, 1948. First published in 1903.

Muller-Ortega, Paul Eduardo. *The Triadic Heart of Śiva.* Albany: State University of New York Press, 1989.

Neumann, Erich. *The Origins and History of Consciousness.* Princeton, N.J.: Princeton University Press, 1995.

Newton, Isaac. *Opticks.* New York: Dover Publications, 1979. First published in 1704.

Omnès, Roland. *Quantum Philosophy.* Princeton, N.J.: Princeton University Press, 1999.

Owen, Richard. *Anatomy of Vertebrates.* London: Longmans and Green, 1866.

———. *On the Nature of Limbs.* London: Jan Van Voorst, 1849.

Piaget, Jean. *The Child's Construction of Reality.* London: Routledge and Kegan Paul, 1955.

Planck, Max. *Wissenschaftliche Selbstbiographie.* Leipzig: Johann Ambrosius Barth, 1948.

Plato, *The Essential Dialogues of Plato.* Translated by Benjamin Jowett. New York: Barnes and Noble, 2005.

Platon. *Sämtliche Werke* (collected works in 4 volumes). Translated into German by Friedrich Schleiermacher. Hamburg, Germany: Rowohlt Taschenbuch Verlag, vols. 1 and 3, 2007; vols. 2 and 4, 2006. First published in 1957.

Polkinghorne, J. C. *The Quantum World.* Middlesex, England: Penguin Books, 1986.

Ricard, Matthieu. *Happiness: A Guide to Developing Life's Most Important Skills.* New York: Little, Brown, 2003.

Ruse, Michael, and E. O. Wilson. "The Approach of Sociobiology: The Evolution of Ethics." In *Religion and the Natural Sciences,* edited by James E. Huchingson, 308–11. New York: Harcourt, Brace, Jovanovich, 1993.

Russell, Bertrand. *History of Western Philosophy.* London: Unwin, 1979. First published in 1946.

Schäfer, Lothar. "Die Quantenwirklichkeit als Manifestation eines kosmischen Bewußtseins und Grundlage für ein neues Bild vom Ursprung des Lebens." In *Theologie und Naturwissenschaft,* edited by F. Vogelsang, 245–52. Bonn, Germany: Evangelische Akademie im Rheinland, 2006.

———. *Em Busca de la Realidad Divina.* Buenos Aires: Lumen Publishing, 2007.

———. *In Search of Divine Reality.* Fayetteville: University of Arkansas Press, 1997.

———. "Nicht-Empirische Wirklichkeit: Die Quantenwirklichkeit als Grundlage der Prä-Darwinistischen Konzeption der Evolution aus der Gesetzlichkeit der Natur." In *Herausforderungen und Grenzen wissenschaftlicher Modelle in Naturwissenschaften und Theologie,* edited by

F. Vogelsang, 169–76. Bonn, Germany: Evangelische Akademie im Rheinland, 2007.

———. "Versteckte Wirklichkeit: Quantentheorie und Transzendenz als Grundlage für ein neues Bild vom Ursprung des Lebens." In *Glaube und Denken. Jahrbuch der Karl-Heim-Gesellschaft 20. Jg. 2007*, edited by Martin Rothgangel and Ulrich Beuttler, 197–222. Frankfurt am Main: Peter Lang GmbH, 2007.

———. *Versteckte Wirklichkeit—Wie uns die Quantenphysik zur Transzendenz führt* (Hidden Reality: How Quantum Physics Will Lead Us to Transcendence). Stuttgart, Germany: Hirzel, 2004.

Schellenberger, Bernardin. *Entdecke dass du glücklich bist* (Discover That You Are Happy). Würzburg, Germany: Echter Verlag, 2006.

Snow, C. P. *The Two Cultures*. Cambridge: Cambridge University Press, 1959.

Stapp, Henry. *Mind, Matter, and Quantum Mechanics*. Berlin: Springer, 1993.

Störig, Hans Joachim. *Kleine Weltgeschichte der Philosophy*. Munich: Droemersche Verlagsanstalt, 1963.

Suzuki, D. T. *Studies in the Lankavatara Sutra*. Delhi: Motilal Banarasidass, 1999.

Teilhard de Chardin, Pierre. *La Place de l'Homme Dans La Nature*. Paris: Edition du Seuil, 1956.

———. *The Phenomenon of Man*. New York: Harper and Brothers, 1959. First published in 1955.

Weikart, Richard. *From Darwin to Hitler*. New York: Palgrave, MacMillan, 2004.

Weinberg, Steven. *Dreams of a Final Theory*. New York: Vintage Books, 1992.

Weizsäcker, Carl Friedrich von. *Die Sterne sind glühende Gaskugeln, und Gott ist gegenwärtig*. Freiburg, Germany: Herder, 1992.

Wheeler, John Archibald, with Kenneth Ford. *Geons, Black Holes and Quantum Foam: A Life in Physics*. New York: Norton and Company, 1998.

PERIODICALS AND PAPERS

Arndt, Markus, Olaf Nairz, Julian Vos-Andreae, Claudia Keller, Gerbrand van der Zouw, and Anton Zeilinger. "Wave-Particle Duality of C60 Molecules." *Nature* 401 (1999): 680–82.

Arsac, Jacques, Mario Beauregard, Raymond Chiao, Freeman Dyson, Bernard d'Espagnat, Nidhal Guessoum, Stanley Klein, Jean Kovalevsky, Dominique Laplane, Mario Molina, Bill Newsome, Pierre Perrier, Lothar Schäfer, Charles Townes, and Trinh Xuan Thuan. "Pour une science sans a priori" [Toward an Open-Minded Science]. *Le Monde,* February 23, 2006.

Cairns, John, Julie Overbaugh, and Stephan Miller. "The Origin of Mutants." *Nature* 335 (1988): 142–45.

Chopra, Deepak. "Meditation Techniques Demonstrated by Deepak Chopra." *The Dr. Oz Show.* http://www.doctoroz.com/videos/deepak-chopra-meditation.

Deacon, T. "Three Levels of Emergent Phenomena." Paper presented to the Science and the Spiritual Quest Boston Conference, October 21–23, 2001.

Denton, Michael J., Craig J. Marshall, and Michael Legge. "The Protein Folds as Platonic Forms: New Support for the Pre-Darwinian Conception of Evolution by Natural Law." *Journal of Theoretical Biology* 219 (2002): 325–42.

Dürr, S., T. Nonn, and G. Rempe. "Origin of Quantum-Mechanical Complementarity Probed by a 'Which-Way' Experiment in an Atom Interferometer." *Nature* 395 (1998): 33–37.

Gähler, Roland, and Anton Zeilinger. "Wave-Optical Experiments with Very Cold Neutrons." *American Journal of Physics* 59 (1991): 316–24.

Jiang, X. M. Cao, B. Teppen, S. Q. Newton, and L. Schäfer. "Predictions of Protein Backbone Structural Parameters from First Principles: Systematic Comparisons of Calculated N-C(α)-C' Angles with High-Resolution Protein Crystallographic Results." *Journal of Physical Chemistry* 99 (1995): 10521–25.

Joseph, Rhawn. "Extinction, Metamorphosis, Evolutionary Apoptosis, and Genetically Programmed Species Mass Death." *Journal of Cosmology* 2 (2009): 235–55.

Klein, Stanley A. "A Dialogue on the Relevance of Quantum Theory to Religion." *Zygon: Journal of Religion and Science* 41 (2006): 567–72.

Maslow, Abraham H. "A Theory of Human Motivation." *Psychological Review* 50 (1943): 370–96.

Peres, Asher. "What Is a State Vector?" *American Journal of Physics* 52 (1984): 644–50.

Pfau, T., S. Spälter, Ch. Kurtsiefer, C. R. Ekstrom, and J. Mlynek. "Loss of Spatial Coherence by a Single Spontaneous Emission. *Physical Review Letters* 73 (1994): 1223–26.

Pollack, Robert. "The Unknown, the Unknowable, and Free Will as a Religious Obligation." Paper presented to the Science and the Spiritual Quest Boston Conference, October 21–23, 2001.

Rezeq, Moh'd, Jason Pitters, and Robert Wolkow. "Tungsten Nanotip Fabrication by Spatially Controlled Field-Assisted Reaction with Nitrogen." *Journal of Chemical Physics* 124 (2006): 204716-1–204716-6.

Rueckner, Wolfgang, and Paul Titcomb, "A Lecture Demonstration of Single Photon Interference." *American Journal of Physics* 64 (1996): 184–88.

Ruse, Michael, "The Bad Smell of Anti-Reductionism." *Research News and Opportunity in Science and Theology* 1, no. 9 (2001): 27.

———. "The Confessions of a Skeptic." *Research News and Opportunity in Science and Theology* 1, no. 6 (2001): 20.

Schäfer, Lothar. "Die Bedeutung der Quantenwirklichkeit für das Verständnis lebender Systeme." Lecture presented at the University of Freiburg, April 27, 2010. http://www.auditorium-netzwerk.de/Neuerscheinungen -2010/Neuerscheinungen-Juni:::6424_7246.html.

———. "Die Quantenphysik und die Philosophia Perennis." *Grenzgebiete der Wissenschaften* 60 (2011): 99–123, 195–219.

———. "Electron Diffraction as a Tool of Structural Chemistry." *Applied Spectroscopy,* 1976, 30, 123–49.

———. "The Gradient Revolution in Structural Chemistry: The Significance of Local Molecular Geometries and the Efficacy of Joint Quantum Mechanical and Experimental Techniques." *Journal of Molecular Structure* 100 (1983): 51–73.

———. "Nonempirical Reality: Transcending the Physical and Spiritual in the Order of the One." *Zygon: Journal of Religion and Science* 43 (2008): 329–52.

———. "Paraklase der Weltsicht—Paraklase der Gottessicht. Wie Umwälzungen in den Naturwissenschaften globale, politische, soziale und religiöse Umwälzungen anzeigen und nach sich ziehen." *Grenzgebiete der Wissenschaft* 58 (2009): 3–48.

———. "Quantum Reality, the Emergence of Complex Order from Virtual States, and the Importance of Consciousness in the Universe." *Zygon: Journal of Religion and Science* 41 (2006): 505–32.

―――. "Quantum Reality and Evolution Theory." *Journal of Cosmology* 3 (2009): 547–57.

―――. "A Response to Carl Helrich: The Limitations and Promise of Quantum Theory." *Zygon: Journal of Religion and Science* 41 (2006): 583–92.

―――. "A Response to Erwin Laszlo: Quantum and Consciousness." *Zygon: Journal of Religion and Science* 41 (2006): 573–82.

―――. "A Response to Stanley Klein: A Dialogue on the Relevance of Quantum Theory to Religion." *Zygon: Journal of Religion and Science* 41 (2006): 593–98.

Schäfer, Lothar, Diogo Valadas Ponte, and Sisir Roy, "Quantenwirklichkeit und Weltethos. Zur Begründung der Ethik in der Ordnung des Kosmos." *Ethica* 17 (2009): 11–54.

―――. "Quantum Reality and Ethos: A Thought Experiment Regarding the Foundation of Ethics in Cosmic Order." *Zygon: Journal of Religion and Science* 44 (2009): 265–87.

Schäfer, Lothar, M. Cao, and M. J. Meadows. "Predictions of Protein Backbone Bond Distances and Angles from First Principles." *Biopolymers* 35 (1995): 603–6.

Schäfer, L., C. Van Alsenoy, and J. N. Scarsdale. "Ab Initio Studies of Structural Features Not Easily Amenable to Experiment. 23. Molecular Structures and Conformational Analysis of the Dipeptide N-acetyl-N'-methyl Glycyl Amide and the Significance of Local Geometries for Peptide Structures." *Journal of Chemical Physics* 76 (1982): 1439–44.

Shoemaker, Stuart D. "English Translation of the Gospel of Thomas." http://www.earlychristianwritings.com/thomas.html.

Villars, C. N. "Microphysical Objects as 'Potentiality Waves.'" *European Journal of Physics* 8 (1987): 148–49.

―――. "Observables, States, and Measurements in Quantum Physics." *European Journal of Physics* 5 (1984): 177–83.

Woese, Carl. "On the Evolution of Cells." *Proceedings of the National Academy of Science* 99 (2002): 8742–47.

Index

Absolute idealism, 26–27
Absorption intensity, 264
Accidental human in disconnected
    world, 165–7
Actualizing our potential, 1
Adaptation, evolution by, 157–60
Aggression, xxi
Alternative medicine, 201–2
Altruism, 175
Amoral robots, human machines as,
    183–7
Amplitudes, 223
    squared, 250
Anaximander of Miletus, 117–18
Animist tradition, 205
Aporia, 189
Archetypes, 24–25
Arete, 168–9
Aristotle, 21–22, 173
Armstrong, Louis, 217
Arndt, Markus, 235
Ataraxia, 173
Atheism, xx–xxi
Atomic orbitals, *254*
Atoms, 12–13, 33, 221–2
    cosmic consciousness and,
        102–5
    oxidized, 269
    planetary model of, 57
    reduced, 269
Augustine of Hippo, 23, 98, 99
Aurobindo, Sri, 95

Bacterial infection, xxiii–xxiv
Base, 133

Bauer, Joachim, 139, 145, 146, 147,
    161, 174
Being, nonmaterial waves as
    principles of, 11–14
Bell, John Stewart, 89
Bell-type experiments, 89–90
Bending of waves, *225*
Berkeley, George, 109
Bestâmi, Bâjewîd, 195, 216
Bible, 12, 27
Big Bang, xi, xiv
Biology, paradox of, 203–4
Bohm, David, 27, 47, 75, 76–77, 83
Böhme, Jakob, 215
Bohr, Niels, 8, 16, 47, 57, 72, 266–8
Boiling point, 108
Boyle, Robert, 184
Brain, xiv
    language and, 6–7
Braque, Georges, 113
Bruno, Giordano, 86–87
Buber, Martin, 216
Buddhist philosophy, 24
Bullets passing through double slit,
    230

Cairns, John, 147
Categorical imperative, 190
Causality, principle of, 110
Chemical bond, 258
Chemical reactions, virtual states in,
    66–68
Chemistry, xxi
Chopra, Deepak, xi–xvii, 78, 159
Classical physics, 19

Coherence, 82, 241
Coincidence, 111
Collapse of wave function, 241
Collective unconscious, 17, 25
Complementarity, 42, 240–1
Consciousness, xiv, 105, 121
    cosmic, *see* Cosmic consciousness
    as cosmic property, 95–124
    creating reality, 51–52
    holistic aspects of, 121–2·
    integrative, 199–201
    mutation of our, 29–30
Constructive interference, 224
Contradiction, epistemological,
    206–7
Cosmic consciousness, 96
    atoms and, 102–5
    reveals itself, 96–101
    wholeness of universe and, 105–7
Cosmic field, 177
Cosmic nature
    of inner images, 16–18
    of morality, 178
Cosmic potential, 1–30
    in you, 91–94
Cosmic potentiality, 204–5
    mindlike aspects of, 18–20
Cosmic property, consciousness as,
    95–124
Cosmic virtual state actualization
    (CVSA), 143–4
Creation, expressed potential of, xii
Cushing, James, 266–7
Cutsinger, James S., 215
CVSA (cosmic virtual state
    actualization), 143–4

Dalai Lama, 173
Dalton, John, 117
Darwin, Charles, 128–9
Da Vinci, Leonardo, xii, 1
Dawkins, Richard, 132–3, 145, 146,
    147
Deacon, Terrence, 139

Decoherence, 82
Degenerate states, 68
Democritus, 117
Density, probability, 59
Denton, Michael J., 130–1, 137, 151,
    152
Depurination, 133
Derain, André, 113
Descartes, René, xxiv–xxv, 30, 184
d'Espagnat, Bernard, 89–90
Destructive interference, 224
Diffraction
    electron, *see* Electron diffraction
    of waves, *225*
    by slit, *227*
Dimensionless numbers, 237
Dimensions, 237
Dirac, Paul Adrien Maurice, 136
Disconnected world, accidental
    human in, 165–7
Divine, thoughts as expression of
    the, 22
DNA, 138
Double slit, 226
    bullets passing through, *230*
    electron diffraction at, *232*
    interference pattern at, *227*
        with monochromatic light, *228*
    wavelike field with, *234*
Duality, 184
    concept of, 42
Dürr, Hans-Peter, 11, 13, 19, 83, 84,
    87, 101, 160, 180, 192, 235

Earth, xii
Eccles, John C., 127
Eckhart, Meister, 30, 209–13
Eddington, Sir Arthur Stanley, 14–
    15, 19, 100, 102–5, 123, 198
Einstein, Albert, xi, xii, xvi, 8–9, 113
EL (entity of life), 64–66
Electric charge, 245
Electron, 245
Electron diffraction, 232

at double slit, *232*
Electronic states, 260–1
Electrons, 12–13, 229
Elementary thoughts, 35
Empirical world, 135
Empty states, 57
Endo, J., *232*
Energy, 50–51, 51, 177, 246
  equivalence of matter and
    potentiality and, 50–51
  psychic, 177
Energy holes, 260
Energy ladders, vibrational, *262*
Energy level diagram, *250*
Energy quanta, 129
Enlightenment, physics of, xix–xxv,
  209–13
Entity, 40
Entity of life (EL), 64–66
Epicurus, 173
Epistemological contradiction, 206–7
Epistemology, 12
Equilibrium positions, 260
ET abbreviation, 41–44, 48, 51–52,
  84, 179–80
Ethics, quantum view of, 181–2
Ethos, world, 164–94
Eudaimonia, 173, 187
Evil in world, 191–4
Evolution, 149
  by adaptation, 157–60
  of life in holistic universe, 144–7
  by natural law, 151–2
  needing quantum selection,
    127–63
Evolving mind, 195–217
Exawa, H., *232*
Excited states, 261
Existence, inner sense of, 176, 179–81
Extended substance, 184

Faulstich, Joachim, 201–2, 208
Feuerstein, Georg, 116
Fields of ghosts, 10

Filled orbitals, 258
Fischbeck, Hans-Jürgen, 101
Ford, Henry, 113
Ford, Kenneth, 54
Forms
  mathematical, *see* Mathematical
    forms
  waveforms, 9
Franck-Condon principle, 264
Frawley, David, 116
Freedom, 182
Free will, 182
Frequency, 223
Freud, Sigmund, 113
Fulfillment, xix
Fundamentalists, religious, 82

Gähler, Roland, 235
Galileo, 30
Gebser, Jean, 29, 115, 196, 199–200
Gene(s), 136–7
  God, 153
  quantum selection and, 132–7
Ghosts, fields of, 10
Giving-not-losing principle of
    potentiality, 98
God, 2, 4, 23, 30, 34–35, 78, 174, 210,
    215
  integrative, 216–7
God Gene, 153
Good will, 193
Goser, Karl, 150
Gospel of Thomas, 28
Gould, Stephen Jay, 147
Gravity, xi, 2
Ground state, 261

Haftmann, Werner, 115
Hamlet, xiii, 1
Happiness, xix, 172–3
Harré, Rom, 267–8
Hauskeller, Michael, 164, 168, 169,
    170, 173
Heat, 8

Hegel, Georg Wilhelm, 26–27
Heisenberg, Werner, 11, 33, 47, 48,
    259
Hemoglobin, 129
Herbert, Nick, 83
Hidden variables, 9
Higgs field, 47, 236
Highest good, 173
Hirschberger, Johannes, 22, 87, 118
Holbach, Paul-Henri-Dietrich,
    Baron d', 185
Holism, 81
Holistic aspects of consciousness,
    121–2
Holistic universe, evolution of life in,
    144–7
Hope, 214–215
Hoyle, Fred, 148
H1s state, 259
Human, accidental, in disconnected
    world, 165–7
Human machines as amoral robots,
    183–7
Human potential movement, xvii
Human significance of quantum
    phenomena, 124
Human values in universe, 167–70
Hume, David, 108, 110, 166, 184
Hüther, Gerald, 3, 16, 25, 212
Hydrogen atom, 57, 61, 248, 249–52
    1.0.0 state of, 259
Hydrogen molecule, *260*

Idealism
    absolute, 26–27
    quantum physics and, 23–28
Idealists, 23
Images, 5
Imperative, categorical, 190
Incomplete information, 242
Indeterminacy, quantum, 231, 241–2
Indistinguishability, 17
Indivisible wholeness, 9
    reality as, 75–94

Infection, bacterial, xxiii–xxiv
Infinite potential, xix
Information, 44–45, 97, 268
    incomplete, 242
Infrared light, 246
Inner images
    cosmic nature of, 16–18
    potential and, 3
Inner sense of existence, 176, 179–81
Integrative consciousness, 199–201
Integrative God, 216–7
Intelligence, 1
Intensity distribution, 236
Interactions, 52
    observation, 268
Interconnectedness, 10–11
Interference
    constructive, 224
    destructive, 224
    single-particle, 221
    of waves, 39, *224*
Interference patterns, 237
    at double slit, *227*
        with monochromatic light, *228*
Interference phenomena, 247
Interferences of wave functions, 13
Internet, 226
Invisible part of reality, xxii
Iso-density surfaces of probability
    distributions, *254, 255, 257*
Is-ought fallacy, 166

James, William, 20
Jeans, Sir James Hopwood, 19,
    135–6
Jesus, 28, 156, 190, 191
Joseph, Rhawn, 146, 148, 149, 150
Joyce, James, xi, 114
Jumps, quantum, 58, 68–70, 132,
    200
Jung, Carl Gustav, 3, 17, 24–25,
    111–13, 122, 183, 188, 191, 192–3,
    194, 198
    psychic energy and, 177

Kafatos, Menas, 27, 89–91, 106, 157
Kafka, Franz, 114
Kak, Subhash, 116
Kandinsky, Wassily, 114
Kant, Immanuel, 30, 173–4, 188, 190, 267
    quantum view of ethics and, 181–2
Karma, 172
Kawasaki, T., *232*
Kindness, 174
Klein, Stanley, 170
Kopper, Joachim, 210–11
Küng, Hans, 184–5, 214

La Mettrie, Julien Offroy de, 184–5
Lancaster, Brian, 27, 157
Language, xxii
    brain and, 6–7
Laszlo, Ervin, 192
Legge, Michael, 151
Leucippus, 117
Life, xi, 149
    evolution of, in holistic universe, 144–7
Light, 247
    infrared, 246
    monochromatic, 246
    speed of, 88
    ultraviolet, 246
    virtual states in interactions of molecules with, 68–70
    visible, 246
    white, 247
Lipton, Bruce, 160, 174
Living in reality, 125–217
Llinás, Rodolfo, 185–6
Localization, 45–46
Logic, 12
Logos, 12
Love, 1

Maine de Biran, 176, 181
Manthey, David, 252, *254*
Marshall, Craig J., 151

Maslow, Abraham, 92–93, 193
Mass, 51
Materialism, 7, 33–53, 74, 184
    birth of, 33–34
Material particles, 221
Material world, 33–53
Mathematical forms, 4, 5
    matter turning into, 36
    power of, 2–6
Matisse, Henri, 113
Matsuda, T., *232*
Matter, 34
    equivalence of energy and potentiality and, 50–51
    metamorphosis of, 233–6
    turning into mathematical forms, 36
Matter-form philosophy, 22
Maxima, 223
    probability, 256
Maxwell, James Clerk, xii
Meaning, universe revealing its, 203–9
Medicine
    alternative, 201–2
    Western, 201–2
Metamorphosis, 148
    of matter, 233–6
Midas, King, 100
Miller, Kenneth, 129, 131
Miller, Stephan, 147
Mind, xiii, xiv, 2–3, 104
    evolving, 195–217
    universe and, 5–6, 107–11
Mind level, 2
Mindlike aspects
    of cosmic potentiality, 18–20
    of virtual states, 70–72
Mindness, 185
Mind-stuff, 105
Minima, 223
Molecular virtual states, 258–9
Molecules
    in quantum states, 261–3

Molecules *(cont.)*
  virtual states in interactions of,
    with light, 68–70
Money, quantum, 246–7
Monochromatic light, 246
  interference pattern at double slit
    with, *228*
Monod, Jacques, 99–100, 128, 132,
    142, 150, 166, 203–7
Moore, G. E., 167
Morality, 188
  cosmic nature of, 178
  quantum phenomena and, 187–91
Moral responsibility, personal
    identity and, 179
Moral rules, 175
Moses, 190
Muller-Ortega, Paul Eduardo,
    119–20
Multichannel devices, 229
Mutations, randomness of, 132–3

Nadeau, Robert, 27, 89–91, 106, 157
Natural law, evolution by, 151–2
Nature, objectivity of, 206
Neumann, John von, 51
Neurology, 102
Newton, Isaac, xii, 34
Newtonian world, xii
Noble gases, 68
Nodal planes, 256
Nonempirical part of reality, 245–70
Nonlocal event, 243
Nonlocality, 88–91
  concept of, 242–4
Nonmaterial waves as principles of
    being, 11–14
Nonseparability, 90
Noumena, 181–2, 267
Nucleotides, 138
Nucleus, 14, 57
Numbers
  theory of, 21
  universe and, 236–8

Objectivity of nature, 206
Object permanence, 109
Observation interactions, 268
Observations, 48
Occupied states, 57
Oesterreicher, Marianne, 83
Omnès, Roland, 114
One, the, 9, 22–23, 28, 49, 77, 84, 87,
    145, 172, 174, 178, 193, 197
Ontology, 12
Orbitals, 59, 251
  atomic, *254*
  filled, 258
Orderliness, xv
Overbaugh, Julie, 147
Owen, Richard, 151
Oxidized atoms, 269
Oxygen, 68

Paradox of biology, 203–4
Parmenides, 34, 37, 73
Particle, 222
Particles, 35
  in a box, example of, 255–8
  characteristics of, 39
  material, 221
  transmaterial, 238
  to waves, 35–40
Particle state, 41, 251
Pascal, Blaise, 214
Perennial philosophy, 116
  as form of synchronicity, 116–121
Peres, Asher, 269
Personal identity, moral
    responsibility and, 179
Perspective, 115
Pfau, T., 235
Philosophy, perennial, *see* Perennial
    philosophy
Photon, 58
Photons, 246–7
Physics, xi, xii, xxi
  classical, 19
  of enlightenment, xix–xxv, 209–13

quantum, *see* Quantum physics
Piaget, Jean, 109, 179
Picasso, Pablo, 113
Pitters, Jason, 270
Planck, Max, xxiii, 113
Planetary model of atom, 57
Plato, 5, 23, 160, 172, 190, 193
Plotinus, 22–23, 28
Polarities, 145
Polkinghorne, J. C., 38
Pollack, Robert, 142
Polymers, 138
Ponte, Diogo Valadas, 165, 187, 266
Popper, Karl, 108
Porpoise, xiv
Potential, xi
    actualizing, 1
    cosmic, 1–30
    expressed, of creation, xii
    infinite, xix
    inner images and, 3
    reality of, 54–74
Potentiality, 46–50, 56, 60, 95
    concept of, 252
    cosmic, 204–5
    equivalence of matter and energy
        and, 50–51
    giving-not-losing principle of, 98
    probability waves and, 239–40
    seeds as centers of, 148–50
    as state of universe, 52–53
    states of, 4
    thoughts as form of, 98
    in universe, 50
    virtual states and, 70–71
    wholeness arising from, 83–87
Potentiality states, 46, 239
Potentiality waves, 47–48, 61
Potential movement, human, xvii
Potential values, 43–44
Primeval stuff, 33
Principle of causality, 110
Probabilities, 97, 237
    transition, 69–70

Probability, zero, *257*
Probability density, 59, 250, 255
Probability distributions, iso-density
    surfaces of, *254, 255, 257*
Probability maxima, 256
Probability patterns, 237
Probability region, zero, 256
Probability waves, 40–44, 44
    potentiality and, 239–40
Proteins, 14, 130
Psychic energy, 177
Purposefulness, 207
Pythagoras, 5, 15, 21

Quanta, xxii
    energy, 129
Quantum, 245
Quantum coherence, wholeness in,
    82–83
Quantum indeterminacy, 231, 241–2
Quantum jumps, 58, 68–70, 132, 200
Quantum money, 246–7
Quantum number, 248
    vibrational, 261
Quantum phenomena
    human significance of, 124
    morality and, 187–91
Quantum physics, xii, xix–xxv
    idealism and, 23–28
    spirituality and, 20–23
Quantum reality, 215
Quantum Reality class, xx, xxiv
Quantum selection, 136, 162
    evolution needing, 127–63
    genes and, 132–7
Quantum states, 57, 248
    molecules in, 261–3
Quantum systems, wholeness in,
    88–91

Randomness, xv
    of mutations, 132–3
Reality, xiv, xvi, 7–9
    aspects of, 171

Reality *(cont.)*
  consciousness creating, 51–52
  domains of, 49
  as indivisible wholeness, 75–94
  invisible part of, xxii
  living in, 125–217
  nature of, 31–124
  nonempirical part of, 245–70
  quantum, 215
  virtual, xii
  virtual states and, 72–74
  words for, 15
Real virtual states, 265
Reason, 201
Redox (reduction-oxidation)
    reactions, 67–68, 269
Reduced atoms, 269
Reductionism, 78–82
Reduction-oxidation (redox)
    reactions, 67–68, 269
Relativity, xi
Religions, 16, 21
Religious fundamentalists, 82
Renaissance, 115
Rezeq, Moh'd, 270
Ricard, Matthieu, 173
Roy, Sisir, 165, 187, 266
Rueckner, Wolfgang, *228,* 247
Ruse, Michael, 154, 155
Russell, Bertrand, xi
Russian Revolution, 114
Rutherford, Ernest, 14

Schäfer, Lothar, xi, xiii, xv, 106–7,
    187
Schellenberger, Bernardin, 190
Schliemann, Heinrich, 115
Schönberg, Arnold, 114, 115
Schrödinger, Erwin, 13, 59, 118,
    249
  in dream, 54–55, 62–66
Schweitzer, Albert, 172
Seeman, Hanne, 28
Selection, quantum, 162

Self, 186
Self-actualization, 93
Single-molecule spectroscopy, 270
Single-particle interference, 221
Slit, diffraction of waves by, *227*
Snow, C. P., 197
Sociobiologists, 154
Socrates, 30, 170, 193
Space, xi
Spanda, 117–20
Spectrometers, 69, 264
Spectroscopy, 247
  single-molecule, 270
Spectrum, 247–8
Speed of light, 88
Spirit, 160
Spirituality, quantum physics and,
    20–23
Squared amplitudes, 250
Standing waves, 251
Stapp, Henry, 136
State(s)
  degenerate, 68
  electronic, 260–1
  empty, 57
  excited, 261
  ground, 261
  H1s, 259
  occupied, 57
  particle, 41, 251
  potentiality, 46, 239
  quantum, *see* Quantum states
  real virtual, 265
  superposition, 240–2
  transition, 134
  twin, 89
  virtual, *see* Virtual states
  wave, 41, 251
State transitions, *250*
Störig, Hans Joachim, 117–18
Stress, 174
Structure of world, 171
Stuff, 104
Subconsciousness, 105

Superposition states, 240–2
Synchronicity, 111–12, 175
    perennial philosophy as form of,
        116–121

Tacit advice, 187
Tacit moral form, 189
Tagore, Rabindranath, xvi
Teilhard de Chardin, Pierre, 29–30,
    144, 162
Teleonomy, 207
Telepathy, 89
Theology, 12
Thinking substance, 184
Thomas, Gospel of, 28
Thoughts
    elementary, 35
    as expression of the divine, 22
    as form of potentiality, 98
Time, xi
Titcomb, Paul, 228, 247
Tonomura, A., 232
Transformation, 148
Transition probabilities, 69–70,
    263–4
Transition state, 134
Transmaterial particles, 238
Triplets of numbers, 252–3
Twin state, 89

Ultraviolet light, 246
Unconscious, collective, 17
Unity of universe, 103
Universe, xi
    holistic, evolution of life in,
        144–7
    human values in, 167–70
    mind and, 5–6, 107–11
    numbers and, 236–8
    outside order of, 128–31
    potentiality as state of, 52–53
    potentiality in, 50
    revealing its meaning, 203–9
    unity of, 103

wholeness of
    cosmic consciousness and, 105–7
    wholesome life in, 171–5

Values, human, in universe, 167–70
van Gogh, Vincent, 175
Varieties of Religious Experience, The
    (James), 20
Vibrational energy, 261
Vibrational energy ladders, 262
Vibrational quantum number, 261
Villars, C. N., 47–48, 190, 268
Virtual, word, 209
Virtual order, 104
Virtual reality, xii
Virtual state actualization (VSA),
    141–2, 151–2
Virtual states, 53, 56, 61–62, 245, 249,
    253
    action of, 54–74
    in chemical reactions, 66–68
    concept of, 252–5
    in interactions with molecules with
        light, 68–70
    mindlike aspects of, 70–72
    molecular, 258–9
    potentiality and, 70–71
    real, 265
    reality and, 72–74
Virtue, 173, 193, 211
Visible light, 246
VSA (virtual state actualization),
    141–2, 151–2

Wave, 222
Waveforms, 9
Wave functions, 249, 255
    collapse of, 241
    interferences of, 13
Wavelength, 223
Wavelets, 35
Wavelike field with double slit, 234
Wave mechanics, 13, 59
Wave-particle duality, 41–42

Waves, 38, 222
    characteristics of, 39
    diffraction of, *225*
    interference of, 39, *224*
    particles to, 35–40
    potentiality, 47–48, 61
    probability, 40–44
    properties of, *223*
    standing, 251
Wave state, 41, 251
Wave train, 222
Weikart, Richard, 156
Weinberg, Steven, 78–81
Weizsäcker, Carl Friedrich von, 190
Western medicine, 201–2
Wheeler, John Archibald, 54
White light, 247
Wholeness, 74, 83
    arising from potentiality, 83–87
    concept of, 76
    indivisible, 9
        reality as, 75–94
    in quantum coherence, 82–83
    in quantum systems, 88–91

of universe
    cosmic consciousness and, 105–7
Wholesome life in wholeness of
    universe, 171–5
Wickramasinghe, Chandra, 148
Wilson, E. O., 154, 155
Woese, Carl, 146
Wolkow, Robert, 270
Work, 51
World
    disconnected, accidental human in,
        165–7
    empirical, 135
    evil in, 191–4
    structure of, 171
World ethos, 164–94
World-Reason, 164
Wright brothers, 113

Zeilinger, Anton, 235
Zeno of Citium, 164, 168, 169, 170
Zero probability, *257*
Zero probability region, 256
Zygote, 148

About the Author

LOTHAR SCHÄFER is the author of *In Search of Divine Reality: Science as a Source of Inspiration* and is a distinguished professor of physical chemistry at the University of Arkansas.